Safety Law

Safety at Work Series*

Volume 1 – Safety Law

Volume 2 – Risk Management

Volume 3 – Occupational Health and Hygiene

Volume 4 – Workplace Safety

---

*These four volumes are available as a single volume, *Safety at Work*, 5th edition.

Safety Law

Volume 1 of the Safety at Work Series

Edited by
John Ridley and John Channing

OXFORD   AUCKLAND   BOSTON   JOHANNESBURG   MELBOURNE   NEW DELHI

Butterworth-Heinemann
Linacre House, Jordan Hill, Oxford OX2 8DP
225 Wildwood Avenue, Woburn, MA 01801-2041
A division of Reed Educational and Professional Publishing Ltd

℞ A member of the Reed Elsevier plc group

First published 1999

© Reed Educational and Professional Publishing Ltd 1999

All rights reserved. No part of this publication
may be reproduced in any material form (including
photocopying or storing in any medium by electronic
means and whether or not transiently or incidentally
to some other use of this publication) without the
written permission of the copyright holder except in
accordance with the provisions of the Copyright,
Designs and Patents Act 1988 or under the terms of a
licence issued by the Copyright Licensing Agency Ltd,
90 Tottenham Court Road, London, England W1P 9HE.
Applications for the copyright holder's written permission
to reproduce any part of this publication should be addressed
to the publishers

British Library Cataloguing in Publication Data
A catalogue record for this book is available from the British Library

Library of Congress Cataloguing in Publication Data
A catalogue record for this book is available from the Library of Congress

ISBN 0 7506 4559 8

Typeset by Genesis Typesetting, Laser Quay, Rochester, Kent
Printed and bound in Great Britain by
Biddles Ltd, Guildford and King's Lynn

Contents

Foreword ix

Preface xi

List of contributors xiii

Introduction 1.1

1   Explaining the law Brenda Watts 1.3
    1.1   Introduction 1.3
    1.2   The incident 1.3
    1.3   Some possible actions arising from the incident 1.3
    1.4   Legal issues of the incident 1.4
    1.5   Criminal and civil law 1.4
    1.6   Branches of law 1.5
    1.7   Law and fact 1.7
    1.8   The courts 1.7
    1.9   Judicial precedent 1.17
    1.10 Court procedure 1.18
    1.11 Identity of court personnel 1.22
    1.12 Industrial Tribunals 1.24
    1.13 European Community Courts 1.25
    1.14 Sources of English law 1.28
    1.15 Legislation 1.29
    1.16 Safety legislation before the Health and Safety at
          Work etc. Act 1.34
    1.17 Safety legislation today 1.35
    1.18 Principles developed by the courts 1.38

2	**Principal health and safety Acts** S. Simpson	1.44
	2.1 The Health and Safety at Work etc. Act 1974	1.44
	2.2 The Factories Act 1961	1.54
	2.3 The Fire Precautions Act 1971	1.54
	2.4 The Mines and Quarries Act 1954–71	1.56
	2.5 The Environmental Protection Act 1990	1.56
	2.6 The Road Traffic Acts 1972–91	1.57
	2.7 The Public Health Act 1936	1.57
	2.8 Petroleum (Consolidation) Act 1928	1.58
	2.9 Activity Centres (Young Persons Safety) Act 1995	1.58
	2.10 Crown premises	1.58
	2.11 Subordinate legislation	1.59
3	**Influences on health and safety** J. R. Ridley	1.62
	3.1 Introduction	1.62
	3.2 The Robens Report	1.62
	3.3 Delegation of law-making powers	1.63
	3.4 Legislative framework for health and safety	1.64
	3.5 Self-regulation	1.65
	3.6 Goal-setting legislation	1.66
	3.7 European Union	1.67
	3.8 European standards	1.70
	3.9 Our social partners	1.71
	3.10 Social expectations	1.72
	3.11 Public expectations	1.72
	3.12 Political influences	1.73
	3.13 Roles in health and safety	1.74
	3.14 Safety culture	1.74
	3.15 Quality culture	1.75
	3.16 No fault liability	1.76
	3.17 Conclusion	1.76
4	**Law of contract** R. W. Hodgin	1.78
	4.1 Contracts	1.78
	4.2 Contracts of employment	1.81
	4.3 Employment legislation	1.81
	4.4 Law of sale	1.84
	4.5 Specialised legislation affecting occupational safety advisers	1.86
5	**Industrial relations law** R. D. Miskin	1.90
	5.1 Introduction	1.90
	5.2 Employment law	1.90
	5.3 Discrimination	1.92
	5.4 Disciplinary procedures	1.99
	5.5 Dismissal	1.102
	5.6 Summary	1.112

6	**Consumer protection** R. G. Lawson	1.**114**
	6.1 Fair conditions of contract	1.**114**
	6.2 A fair quality of goods and services	1.**121**
	6.3 Product safety	1.**122**
	6.4 Product liability	1.**126**
	6.5 Misleading advertising	1.**127**
	6.6 Exclusion clauses	1.**128**
	6.7 Consumer redress	1.**130**
7	**Insurance cover and compensation** A. West	1.**133**
	7.1 Workmen's compensation and the State insurance scheme	1.**133**
	7.2 Employer's liability insurance	1.**136**
	7.3 Public Liability insurance	1.**142**
	7.4 Investigation, negotiation and the quantum of damage	1.**143**
	7.5 General	1.**146**
8	**Civil liability** E. J. Skellett	1.**147**
	8.1 The common law and its development	1.**147**
	8.2 The law of tort	1.**148**
	8.3 Occupier's Liability Acts 1957 and 1984	1.**151**
	8.4 Supply of goods	1.**152**
	8.5 Employer's liability	1.**153**
	8.6 Employer's Liability (Defective Equipment) Act 1969	1.**156**
	8.7 Health and Safety at Work etc. Act 1974	1.**156**
	8.8 Defences to a civil liability claim	1.**156**
	8.9 Volenti non fit injuria	1.**158**
	8.10 Limitation	1.**158**
	8.11 Assessment of damages	1.**159**
	8.12 Fatal accidents	1.**160**
	8.13 'No fault' liability system	1.**161**

Appendix 1 Institution of Occupational Safety and Health   1.**163**

Appendix 2 Reading for Part 1 of the NEBOSH Diploma examination   1.**164**

Appendix 3 List of Abbreviations   1.**165**

Appendix 4 Organizations providing safety information   1.**171**

Appendix 5 List of Statutes, Regulations and Orders   1.**173**

Appendix 6 List of Cases   1.**181**

Appendix 7 Safety at Work Series – Index   1.**186**

Foreword

Frank J. Davies CBE, O St J, *Chairman, Health and Safety Commission*

My forty years experience of working in industry have taught me the importance of health and safety. Even so, since becoming Chairman of the Health and safety Commission (HSC) in October 1993, I have learned more about the extent to which health and safety issues impact upon so much of our economic activity. The humanitarian arguments for health and safety should be enough, but if they are not the economic ones are unanswerable now that health and safety costs British industry between £4 billion and £9 billion a year. Industry cannot afford to overlook these factors and needs to find a way of managing health and safety for its workers and for its businesses.

In his foreword to the third edition of this book my predecessor, Sir John Cullen, commented on the increasing impact of Europe in the field of health and safety, most notably through European Community Directives. We have since found this to be very much so. I believe that the key challenge health and safety now faces is to engage and influence the huge variety of businesses, particularly small businesses, and to help them manage health and safety more effectively. I would add that the public sectors, our largest employers these days, also should look at their management of health and safety to ensure they are doing enough.

Many businesses are willing to meet their legal obligations if given a gentle prompt and the right advice and HSC is very conscious of the importance of having good regulations which are practicable and achievable.

It is, of course, vital and inescapable that an issue as critical as health and safety should be grounded in sound and effective legislation.

This book covers many of these and other important health and safety developments, including environmental and industrial relations law which touch on this area to varying degrees. I welcome the contribution it makes towards the goal of reaching and maintaining effective health and safety policies and practices throughout the workplace.

Preface

Health and safety is not a subject in its own right but is an integration of knowledge and information from a wide spectrum of disciplines. *Safety at Work* reflects this in the range of chapters written by experts and in bringing the benefits of their specialised experiences and knowledge together in a single volume.

While there is a continuing demand for a single volume, many managers and safety practitioners enter the field of safety with some qualifications already gained in an earlier part of their career. Their need is to add to their store of knowledge specific information in a particular sector. Equally, new students of the subject may embark on a course of modular study spread over several years, studying one module at a time. Thus there appears to be a need for each part of *Safety at Work* to be available as a stand-alone volume.

We have met this need by making each part of *Safety at Work* into a separate volume whilst, at the same time, maintaining the cohesion of the complete work. This has required a revision of the presentation of the text and we have introduced a pagination system that is equally suitable for four separate volumes and for a single comprehensive tome. The numbering of pages, figures and tables has been designed so as to be identified with the particular volume but will, when the separate volumes are placed together as a single entity, provide a coherent pagination system.

Each volume, in addition to its contents list and list of contributors, has appendices that contain reference information to all four volumes. Thus the reader will not only have access to the detailed content of the particular volume but also information that will refer him to, and give him an overview of, the wider fields of health and safety that are covered in the other three volumes.

In this way we hope we have kept in perspective the fact that while each volume is a separate part, it is only one part, albeit a vital part, of a much wider spectrum of disciplines that go to make occupational health and safety.

John Ridley
John Channing
October 1998

Contributors

R.W. Hodgin, LL.M.
Senior Lecturer in Law, The University of Birmingham. Consultant, Sennenschein
Dr R.G. Lawson, LL.M., PhD
Consultant in marketing and advertising law
R. D, Miskin
A solicitor
John Ridley, BSc(Eng), CEng, MIMechE, FIOSH, DMS
Stan Simpson, CEng, MIMechE, MIOSH
Eric J. Skellett
A solicitor
Brenda Watts, MA, BA
Barrister, Senior Lecturer, Southampton College of Higher Education
Ashton West, BA(Hons), ACII
General Manager, Iron Trades Insurance Group

Introduction

Laws are necessary for the government and regulation of the affairs and behaviour of individuals and communities for the benefit of all. As societies and communities grow and become more complex, so do the laws and the organisation necessary for the enforcement and administration of them.

The industrial society in which we live has brought particular problems relating to the work situation and concerning the protection of the worker's health and safety, his employment and his right to take 'industrial action'.

This book looks at how laws are administered in the UK and the procedures to be followed in pursuing criminal actions and common law remedies through the courts. It considers various Acts and Statutes that are aimed at safe working in the workplace and also some of the influences that determine the content of new laws. Further, the processes are reviewed by which liabilities for damages due to either injury or faulty product are established and settled.

Chapter 1
Explaining the law
Brenda Watts

1.1 Introduction

To explain the law an imaginary incident at work is used which exemplifies aspects of the operation of our legal system. These issues will be identified and explained with differences of Scottish and Irish law being indicated where they occur.

1.2 The incident

Bertha Duncan, an employee of Hazards Ltd, while at work trips over some wire in a badly lit passageway, used by visitors as well as by employees. The employer notifies the accident in accordance with his statutory obligations. The investigating factory inspector, Instepp, is dissatisfied with some of the conditions at Hazards, so he issues an improvement notice in accordance with the Health and Safety at Work etc. Act 1974 (HSW), requiring adequate lighting in specified work areas.

1.3 Some possible actions arising from the incident

The *inspector*, in his official capacity, may consider a prosecution in the criminal courts where he would have to show a breach of a relevant provision of the safety legislation. The likely result of a successful safety prosecution is a fine, which is intended to be penal. It is not redress for Bertha.

The *employee*, Bertha, has been injured. She will seek money compensation to try to make up for her loss. No doubt she will receive State industrial injury benefit, but this is intended as support against misfortune rather than as full compensation for lost wages, reduced future prospects or pain and suffering. Bertha will therefore look to her employer for compensation. She may have to consider bringing a civil action, and will then seek legal advice (from a solicitor if she has no union to turn to) about claiming compensation (called damages). To succeed,

Bertha must prove that her injury resulted from breach of a legal duty owed to her by Hazards.

For the *employer*, Hazards Ltd, if they wish to dispute the improvement notice, the most immediate legal process will be before an industrial tribunal. The company should, however, be investigating the accident to ensure that they comply with statutory requirements; and also in their own interests, to try to prevent future mishaps and to clarify the facts for their insurance company and for any defence to the factory inspector and/or to Bertha. The company would benefit from reviewing its safety responsibilities to non-employees (third parties) who may come on site. As a company, Hazards Ltd has legal personality; but it is run by people and if the inadequate lighting and slack housekeeping were attributable to the personal neglect of a senior officer (s. 37 HSW), as well as the company being prosecuted, so too might the senior officer.

1.4 Legal issues of the incident

The preceding paragraphs show that it is necessary to consider:

> criminal and civil law,
> the organisation of the courts and court procedure,
> procedure in industrial tribunals, and
> the legal authorities for safety law: legislation and court decisions.

1.5 Criminal and civil law

A *crime* is an offence against the State. Accordingly, in England prosecutions are the responsibility of the Crown Prosecution Service; or, where statute allows, an official such as a factory inspector (ss. 38, 39 HSW). Very rarely may a private person prosecute. In Scotland the police do not prosecute since that responsibility lies with the procurators-fiscal, and ultimately with the Lord Advocate. In Northern Ireland the Director of Public Prosecutions (DPP) initiates prosecutions for more serious offences, and the police for minor cases. The DPP may also conduct prosecutions on behalf of Government Departments in magistrates' courts when requested to do so. The procurators-fiscal, and in England and Northern Ireland the Attorney General acting on behalf of the Crown, may discontinue proceedings; an individual cannot.

Criminal cases in England are heard in the magistrates' courts and in the Crown Court; in Scotland mostly in the Sheriff Court, and in the High Court of Justiciary. In Northern Ireland criminal cases are tried in magistrates' courts and in the Crown Court. In all three countries the more serious criminal cases are heard before a jury, except in Northern Ireland for scheduled offences under the Northern Ireland (Emergency Provisions) Acts of 1978 and 1987.

The burden of proving a criminal charge is on the prosecution; and it must be proved beyond reasonable doubt. However, if, after the incident at Hazards, Instepp prosecutes, alleging breach of, say s. 2 of HSW, then Hazards must show that it was not reasonably practicable for the company to do more than it did to comply (s. 40 HSW). This section puts the burden on the accused to prove, on the balance of probabilities, that he had complied with a practicable or reasonably practicable statutory duty under HSW.

The rules of evidence are stricter in criminal cases, to protect the accused. Only exceptionally is hearsay evidence admissible. In Scotland the requirement of corroboration is stricter than in English law.

The main sanctions of a criminal court are imprisonment and fines. The sanctions are intended as a punishment, to deter and to reform, but not to *compensate* an injured party. A magistrates' court may order compensation to an individual to cover personal injury and damage to property. Such a compensation order is not possible for dependants of the deceased in consequence of his death. At present the upper limit for compensation in the magistrates' court is £5000[1].

A *civil action* is between individuals. One individual initiates proceedings against another and can later decide to settle out of court. Over 90% of accident claims are so settled.

English courts hearing civil actions are the county courts and the High Court; in Scotland the Sheriff Court and the Court of Session. In Northern Ireland the County Court and the High Court deal with civil accident claims. Civil cases rarely have a jury; in personal injury cases only in the most exceptional circumstances.

A civil case must be proved on the balance of probabilities, which has been described as 'a reasonable degree of probability ... more probable than not'. This is a lower standard than the criminal one of beyond reasonable doubt, which a judge may explain to a jury as 'satisfied so that you are sure' of the guilt of the accused.

In civil actions the plaintiff (the pursuer in Scotland) sues the defendant (the defender) for remedies beneficial to him. Often the remedy sought will be damages – that is, financial compensation. Another remedy is an injunction, for example, to prevent a factory committing a noise or pollutant nuisance.

1.6 Branches of law

As English law developed it followed a number of different routes or branches. The diagram in *Figure 1.1* illustrates the main legal sources of English law and some of the branches of English law.

Criminal law is one part of public law. Other branches of public law are constitutional and administrative law, which include the organisation and jurisdiction of the courts and tribunals, and the process of legislation.

Civil law has a number of branches. Most relevant to this book are contract, tort (delict in Scotland) and labour law. A contract is an agreement between parties which is enforceable at law. Most commercial law (for example, insurance) has a basis in contract. A tort is a breach of duty imposed by law and is often called a civil wrong. The two most

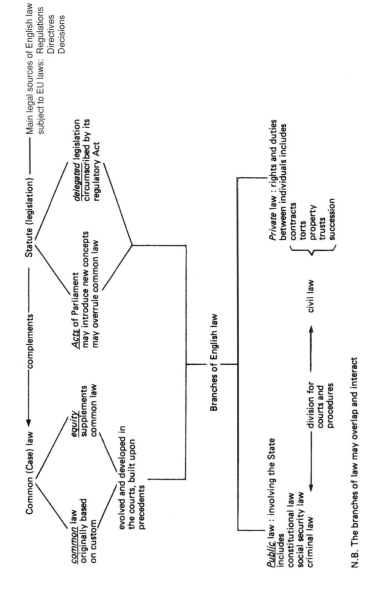

Figure 1.1 Sources and branches of English law

frequently heard of torts are nuisance and trespass. However, the two most relevant to safety law are the torts of negligence and of breach of statutory duty.

The various branches of law may overlap and interact. At Hazards, Bertha has a *contract* of employment with her employer, as has every employee and employer. An important implied term of such contracts is that an employer should take reasonable care for the safety of employees. If Bertha proves that Hazards were in breach of that duty, and that in consequence she suffered injury, Hazards will be liable in the *tort* of negligence. There could be potential *criminal* liability under HSW. Again, Hazards might discipline a foreman, or Bertha's workmates might refuse to work in the conditions, taking the situation into the field of *industrial relations* law.

1.7 Law and fact

It is sometimes necessary to distinguish between questions of law and questions of fact.

A jury will decide only questions of fact. *Questions of fact* are about events or the state of affairs and may be proved by evidence. *Questions of law* seek to discover what the law is, and are determined by legal argument. However, the distinction is not always clear-cut. There are more opportunities to appeal on a question of law than on a question of fact.

Regulation 12 of the Workplace (Health, Safety and Welfare) Regulations 1992 (WHSW) requires an employer (and others, to the extent of their control) to keep, so far as reasonably practicable, every floor in the workplace free from obstructions and from any article which may cause a person to slip, trip or fall. In the Hazards incident Bertha's tripping, her injury, the wire being there, the routine of Hazards, are questions of fact. However, the meaning of 'obstruction', of 'floor', of 'reasonably practicable' are questions of law.

1.8 The courts

1.8.1 First instance: appellate

A court may have *first instance* jurisdiction, which means that it hears cases for the first time; it may have *appellate* jurisdiction which means that a case is heard on appeal; or a court may have both.

1.8.2 Inferior: superior

Inferior courts are limited in their powers: to local jurisdiction, in the seriousness of the cases tried, in the sanctions they may order, and, in England, in the ability to punish for contempt.

In England the superior courts are the House of Lords, the judicial Committee of the Privy Council, and the Supreme Court of Judicature. Magistrates' and county courts are inferior courts.

For Scotland the Sheriff Court is an inferior court while the superior courts are the House of Lords, the Court of Session and the High Court of Justiciary.

In Northern Ireland the superior courts are the House of Lords and the Supreme Court of Judicature of Northern Ireland. The inferior courts are the magistrates' courts and the county courts.

1.8.3 Criminal proceedings – trial on indictment; summary trial

The indictment is the formal document containing the charge(s), and the trial is before a judge and a jury (of 12 in England and N. Ireland, of 15 in Scotland). A summary trial is one without a jury.

The most serious crimes, such as murder, or robbery, must be tried on indictment (or solemn procedure in Scotland). Some offences are triable only summarily (for example, most road traffic offences), others (for example, theft) are triable either way according to their seriousness. Most offences under HSW are triable either way, but in practice are heard summarily.

1.8.4 Representation

A practising lawyer will be a solicitor or a barrister (advocate in Scotland). Traditionally, barristers concentrate on advocacy and provide specialist advice. A qualification for senior judicial appointment is sufficient experience as an advocate. A barrister who has considerable experience and thinks he has attained some distinction may apply to the Lord Chancellor to 'take silk'. A solicitor is likely to be a general legal adviser. Until the Courts and Legal Services Act 1990, a solicitor's right to represent in court was limited to the lower courts. That Act provides for the ending of the barrister's monopoly appearances in the higher courts. Solicitors will be able to appear in the High Court and before juries; and be appointed judges in the High Court. Qualified Fellows of the Institute of Legal Executives now have certain rights of audience, particularly in county courts and tribunals. A party may always defend himself, but there are restrictions on an individual personally conducting a private prosecution in the Crown Court or above (*R. v. George Maxwell Ltd*[2]). There is no general right of private prosecution in Scotland.

1.8.5 An outline of court hierarchy in England

There is a system of courts for hearing civil actions and a system for criminal actions. These are shown diagrammatically in *Figures 1.2* and *1.3*. However, some courts have both civil and criminal jurisdiction.

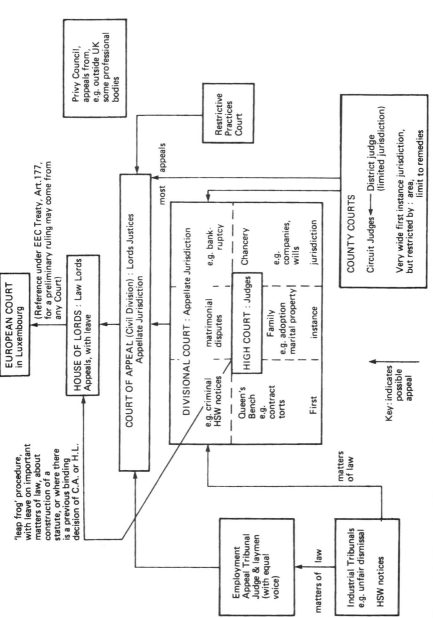

Figure 1.2 The main civil courts in England

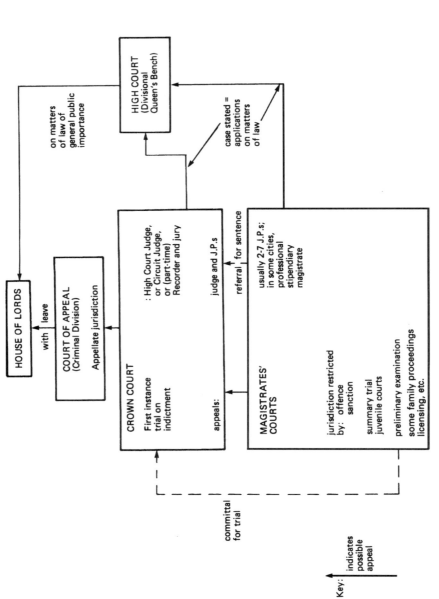

Figure 1.3 The main criminal courts in England

The lowest English courts are the magistrates' courts, which deal mainly with criminal matters; and the county courts, which deal only with civil matters.

Magistrates determine and sentence for many of the less serious offences. They also hold preliminary examinations into other offences to see if the prosecution can show a prima facie case on which the accused may be committed for trial. Serious criminal charges (and some not so serious where the accused has the right to jury trial) are heard on indictment in the Crown Court. The Crown Court also hears appeals from magistrates. For civil cases, the Courts and Legal Services Act increases the jurisdiction of the county courts. All personal injury claims for less than £50 000 will start in the county court; there is no upper limit but county court jurisdiction depends on the complexity of the case. District judges attached to the small claims courts may deal with personal injury cases for less than £5000. More important civil matters, because of the sums involved or legal complexity, will start in the High Court of Justice. The High Court has three divisions:

> Queen's Bench (for contract and torts),
> Chancery (for matters relating to, for instance,
>   land, wills, partnerships and companies),
> Family.

In addition the Queen's Bench Division hears appeals on matters of law:

1 from the magistrates' courts and from the Crown Court on a procedure called 'case stated', and
2 from some tribunals, for example the finding of an industrial tribunal on an enforcement notice under HSW.

It also has some supervisory functions over lower courts and tribunals if they exceed their powers or fail to carry out their functions properly, or at all.

The High Court, the Crown Court and the Court of Appeal are known as the Supreme Court of Judicature.

The Court of Appeal has two divisions: the Civil Division which hears appeals from the county courts and the High Court; and the Criminal Division which hears appeals from the Crown Court. Further appeal, in practice on important matters of law only, lies to the House of Lords from the Court of Appeal and in restricted circumstances from the High Court. The Judicial Committee of the Privy Council is not part of the mainstream judicial system, but hears appeals, from, for instance, the Channel Islands, some Commonwealth countries and some disciplinary bodies.

Since our entry into the European Community, our courts must follow the rulings of the European Court of Justice. On an application from a member country, the European Court will determine the effect of European directives on domestic law. Potentially, the involvement is far-reaching in industrial obligations, including safety.

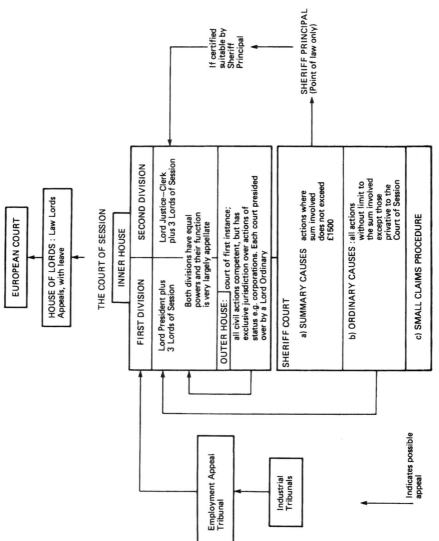

Figure 1.4 The main civil courts in Scotland

1.8.6 Court hierarchy in Scotland

Scotland also has separate but parallel frameworks for the organisation of its civil and criminal courts. These are shown diagrammatically in *Figures 1.4* and *1.5* and are discussed below.

The court most used is the local Sheriff Court which has wide civil and criminal jurisdiction. Civilly it may sit as a court of first instance or as a court of appeal (to the Sheriff Principal from a sheriff's decision). For criminal cases the sheriff sits with a jury for trials on indictment, and alone to deal with less serious offences prosecuted on complaints, when its jurisdiction encompasses that of the restricted district court.

The Court of Session is the superior civil court. The Outer House, sometimes sitting with a jury, has original jurisdiction; the Inner House hears appeals from the Sheriff Court and from the Outer House. Matters of law may be referred to the Inner House for interpretation, and it also hears appeals on matters of law from some committees and tribunals, such as decisions on HSW enforcement notices. Appeal from

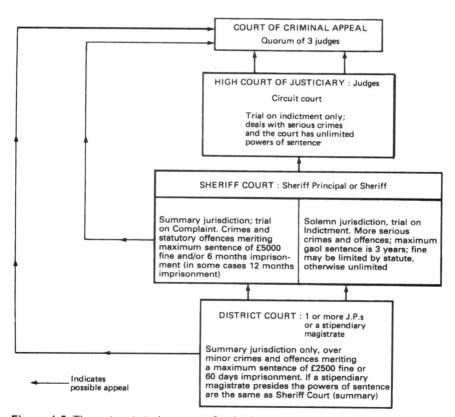

Figure 1.5 The main criminal courts in Scotland

the Inner House is to the House of Lords. For criminal cases the final court of appeal is the High Court of Justiciary, with three or more judges. When sitting with one judge and a jury it is a court of first instance, having exclusive jurisdiction in the most serious criminal matters and unrestricted powers of sentencing. The High Court of Justiciary hears appeals from the first instance courts but only on matters of law in cases tried summarily in the Sheriff Court and the district courts. The judges of the High Court are the same persons as the judges of the Court of Session. They have different titles and wear different robes.

1.8.7 Court hierarchy in Northern Ireland

The hierarchy of courts in Northern Ireland is different from that for the English courts and is shown in *Figures 1.6* and *1.7*.

Most criminal charges are heard in the magistrates' courts. Magistrates try summary accusations or indictable offences being dealt with summarily. They also undertake a preliminary examination of a case to be heard in the Crown Court on indictment (committal proceedings).

Following trial in a magistrates' court, the defendant may appeal to the county court; or, on matters of law only, by way of 'case stated' to the Court of Appeal. The prosecution may appeal only to the Court of Appeal and only on a matter of law by way of 'case stated'. Trial on indictment, for more serious offences, is in the Crown Court, before a judge and jury (except for scheduled offences under the emergency legislation when cases are heard before a judge alone).

Appeal from the Crown Court is to the Court of Appeal. The defendant needs leave unless he is appealing only on a matter of law. The prosecution may refer a matter of law to the Court of Appeal, but this will not affect an acquittal. Final appeal by either side is to the House of Lords, but only with leave and only on matters of law of general public importance. Some civil proceedings take place in a magistrates' court before a resident magistrate (RM). County courts have a wider and almost exclusive civil first instance jurisdiction. The procedure is less formal than in English county courts. Appeal from a County Court is to the High Court for a rehearing, or to the Court of Appeal on a matter of law only.

The High Court has unlimited civil jurisdiction. Appeal by way of rehearing is to the Court of Appeal; or in exceptional circumstances on important matters of law, direct to the House of Lords. Appeal from the Court of Appeal to the House of Lords is possible on matters of law only and with leave.

The Divisional Court hears application for judicial review and habeas corpus in contrast to the wider jurisdiction on 'case stated' of the English court and the English Divisional Courts for Chancery and Family.

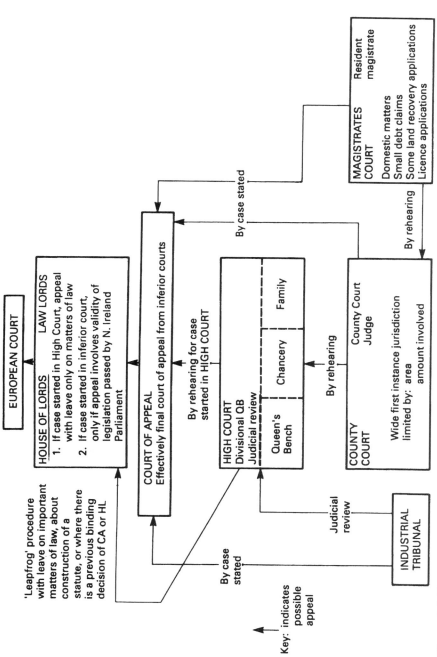

Figure 1.6 The civil courts in Northern Ireland

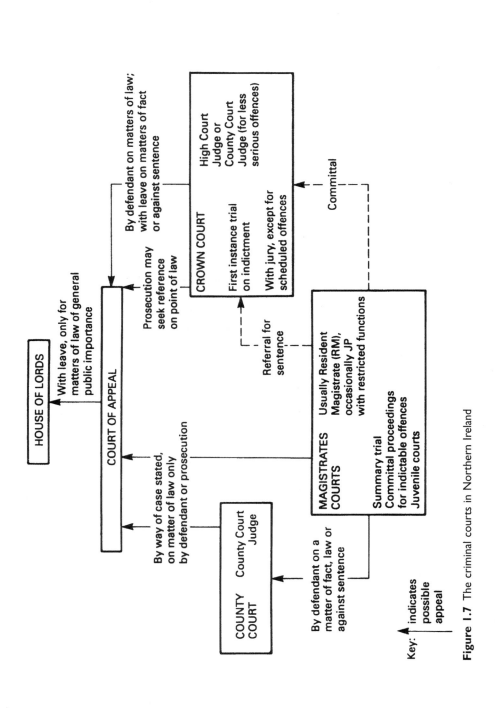

Figure 1.7 The criminal courts in Northern Ireland

1.9 Judicial precedent

Previous court decisions are looked to for guidance. English law has developed a strong doctrine of judicial precedent (sometimes referred to as *stare decesis* – let the decision stand). Some decisions (precedents) **must** be followed in a subsequent case. Other precedents are only persuasive. To operate the doctrine of judicial precedent it is necessary to know:

1 the legal principle of a judgement, and
2 when a decision is binding and when persuasive.

Higher courts bind lower courts, and superior courts usually follow their own previous decisions unless there is good reason to depart from them. Only since 1966 has departure been possible for the House of Lords, and the Civil Division of the Court of Appeal is not expected to depart from its own properly made previous decisions. The Criminal Division has more latitude because the liberty of the accused may be affected.

Decisions of the superior courts which are not binding are *persuasive*, judicial decisions of other common law countries or from the Judicial Committee of the Privy Council (see below: 1.14, para. 2) are also persuasive. The judgements of inferior courts are mostly on questions of fact and are not strict precedents. Decisions of the Court of Justice of the European Communities bind English courts on the interpretation of EC legislation.

The legal principle of a judgement, the actual findings on the particular facts, is called the *ratio decidendi*. Any other comments, such as what the likely outcome would have been had the facts been different, or reference to law not directly relevant, are persuasive but not binding. They are called *obiter dicta* – 'comments by the way'. The *obiter dicta* can be so persuasive that they are incorporated into later judgements and become part of the *ratio decidendi*. This happened to the *dicta* in the famous negligence case of *Hedley Byrne* v. *Heller & Partners*[3] *(see below: 1.18, para. 4)*. Also, *obiter* is any dissenting judgement.

A precedent can bind only on similar facts. A court may *distinguish* the facts in a present case from those in an earlier case so that a precedent may not apply. A previous decision which has been distinguished may still be *persuasive*. An appeal court may *approve* or *disapprove* a precedent. A higher court may *overrule* a precedent, i.e. overturn a principle (though not the actual decision) of a lower court in a different earlier case. If a decision of a lower court is taken to appeal, the higher court will confirm or *reverse* the specific original decision.

The English doctrine of judicial precedent has evolved to give certainty and impartiality to a legal system relying upon case law decisions. Other advantages of the doctrine are the range of cases available and the practical information therein is said to provide flexibility for application to new circumstances and at the same time detailed guidance. Criticisms of the doctrine are that it is not always easy to discover the *ratio decidendi* of a judgement. One way in which a court may avoid a previous decision is to hold that it is *dicta* and not *ratio*. Other criticisms are that the doctrine leads to rigid compliance in a later case unless the previous decision can be distinguished; and that trying to avoid or distinguish a precedent can lead to legal deviousness. The doctrine of binding judicial precedent applies similarly in N. Ireland. In Scotland precedent is important, but there is also emphasis on principle. The European Court of Justice regards precedents but is not bound by them.

For the doctrine of precedent to operate there must be reliable law reporting. Important judgements are published in the Weekly Law Reports (WLR), some of which are selected for the Law Reports. Another important series is the All England Reports (All ER). Important Scottish cases are reported in Sessions Cases (SC) and Scots Law Times (SLT). In N. Ireland the two main series of law reports are the Northern Ireland Law Reports (NI) and the Northern Ireland Judgements Bulletin (NIJB), sometimes called the Bluebook. There are various specialist law reports, to which reference may be made when considering safety cases. A list of their abbreviations is published in *Current Law*[4] which also summarises current developments and current accident awards.

Legal terminology in Law Reports includes abbreviations such as LJ (Lord Justice), MR (Master of the Rolls), per Mr Justice Smith (meaning 'statement by'); *per curiam* means statement by all the court; *per incuriam* means failure to apply a relevant point of law.

A decision of a higher court is a precedent, even though it is not reported in a law report. As well as written law reports, there are computerised data bases. An important example is Lexis[5], which includes unreported judgements of the Civil Division of the Court of Appeal. This very useful development may also accentuate a practical problem of the doctrine of judicial precedent. The volume of cases which may be cited may unnecessarily complicate a submission and lengthen legal hearings. This danger has been recognised in the House of Lords[6].

1.10 Court procedure

1.10.1 Stages in the proceedings

English, Irish and Scottish law follow an 'adversary' system, in which each side develops its cases and answers the contentions of the other. The judge's functions are to ensure that the correct procedures are followed, to clarify ambiguities, and to decide the issue. He may question, but he should not 'come down into the arena' and enter into argument.

An indication of the possible proceedings that could arise following an accident to an employee at work are shown in *Figure 1.8* and considered below.

Referring to the incident, should criminal proceedings be instituted against Hazards, in England and Wales any information stating the salient facts is laid before a magistrate.

Section 38 of HSW requires this to be by an inspector or by or with the consent of the Director of Public Prosecutions. The magistrate will issue a summons to bring the defendant before the court, and this would be served on Hazards at their registered office. Since a company has no

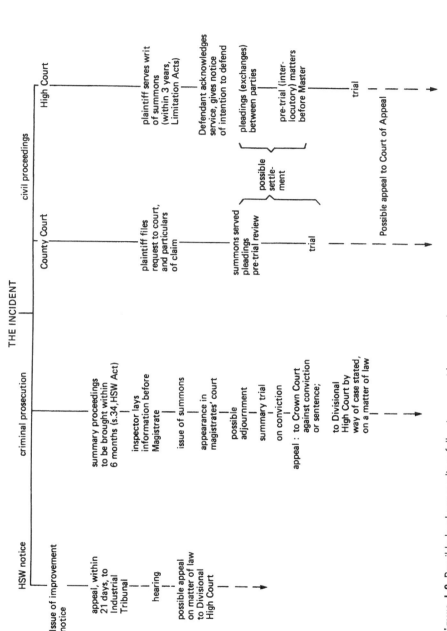

Figure 1.8 Possible legal proceedings following an accident at work

physical existence, and therefore cannot represent itself, it would act through a solicitor or barrister.

In Scotland offences are reported to the local procurator-fiscal who decides whether to prosecute (and in what form when offences are triable either way). With serious cases he would consult with the Crown Office. If there is to be a summary trial a complaint is served on the accused stating the details of the charge.

Most HSW prosecutions are heard summarily, and then trial may commence when the accused is before the magistrates (in England and N. Ireland) or the sheriff (in Scotland). In England and N. Ireland, if the trial is to be on indictment, the magistrates will sit as examining justices to see if there is a case to answer before committing the accused for trial at the Crown Court. A magistrate may issue a witness summons and a procurator fiscal a citation if it appears that a witness will not attend voluntarily.

In a civil claim in the High Court or Court of Session Bertha Duncan, the plaintiff (pursuer), starts her action by obtaining a writ of summons and then serving this on Hazards Ltd. Hazards would consult their solicitors who would acknowledge service and indicate whether they intend to contest proceedings (if they do not, there may be judgement in default).

Then come the pleadings when the plaintiff details the grounds of her claim and the damages she is claiming; and the defendant replies to the specific allegations.

Before trial each side must disclose to the other the existence of documents relevant to its case. The other side is allowed to inspect documents which are not privileged. An important ground of privilege is the protection of communication between a party and his legal advisers. In 1979 the House of Lords in *Waugh* v. *British Railways Board*[7] held that legal advice must be the dominant purpose of a document for it to be privileged. In this case disclosure was ordered of the report of a works accident, incorporating witnesses' statements, which while intended to establish the cause of the accident was intended also for the Board's solicitors.

An order (subpoena) requiring the attendance of a witness may be obtained. In N. Ireland witnesses may remain in court during the hearing of evidence, unlike England.

Proceedings in the inferior courts are similar to those in the High Court and Court of Session, but quicker, cheaper and more under the direction of the court administrators.

Usually a criminal case is decided before a related civil hearing comes on. The Civil Evidence Act 1968 (1971 for N. Ireland) allows a conviction to be used in subsequent civil proceedings. The conviction and the intention to rely on it must be set out in the formal civil pleadings. If this happened with Hazards then it would be for the company to plead and to prove (on the balance of probabilities) that the conviction is irrelevant or was erroneous. In civil personal injury claims, settlement rather than court trial is a likely outcome, under the guidance of insurers.

If Bertha Duncan (*née* Smith) is suing in Scotland her case is referred to as *Smith (or Duncan)* v. *Hazards Ltd*, though for brevity it may be quoted as *Duncan* v. *Hazards Ltd*. The latter is also the English and N. Ireland practice (in speech the case is referred to as Duncan *and* Hazards Ltd).

On appeal, the party appealing, who may have been the defendant in the earlier trial, may be called the appellant and the other party the respondent.

1.10.2 The burden of proof

The phrase 'burden of proof' may be used in two senses. The underlying burden is on the prosecution or plaintiff to prove liability, sometimes called the 'legal' or 'final' burden of proof. However, during the trial the defendant may, for example, dispute evidence or argue a defence. The 'evidentiary' burden of proof then shifts to the defendant, but will shift back to the prosecution if it wishes to dispute that evidence of the defendants. The defendant's evidentiary burden of proof is on the balance of probabilities, even in a criminal trial.

1.10.3 The accused

With a criminal prosecution, normally the accused must attend court to answer the allegation(s) put to him. However, with offences triable only summarily (before magistrates) carrying a maximum penalty of three months or less[8], the accused may plead guilty in writing. The accused must answer every count (offence) alleged. Any acknowledgement of guilt must be unmistakeable and made freely without undue pressure from counsel or the court. If a guilty plea is made in error, it may be withdrawn at any time before sentence. A plea of not guilty may be changed during the trial with the judge's leave. It is possible for plea arrangements to be made between prosecuting and defence counsel where a plea of guilty to a lesser charge is accepted in return for the prosecution not proceeding with a more serious charge; or for a guilty plea to allow consideration of a sentence concession.

The accused has a right to silence, but since the Criminal Justice and Public Order Act 1994 there are greater risks in maintaining that position[9]. However, there can be no conviction on silence alone. There are statutory restrictions on questioning the accused about any criminal past and bad character[10]; and there are strict rules as to the admission as evidence of confessions of guilt[11].

1.10.4 Witnesses

The function of a witness is to inform the judge or the jury of facts, not opinions, unless the witness is called as an expert witness. Most people can be compelled to be witnesses[12]. Failure to comply with a witness order is contempt of court. A witness will be questioned by counsel who called him/her and may then be cross-examined by counsel for the other side. Counsel who called the witness may re-examine but may not raise new issues. 'Leading' questions (a question suggesting an answer) may not be asked. A witness cannot be compelled to answer a question which may incriminate him/her. A witness's evidence is usually given orally in open court, but in certain circumstances (e.g. illness) evidence is allowed by affidavit (a sworn written statement).

Expert evidence is opinion evidence on a technical point(s). Opinion evidence is admissible from an expert but not from an ordinary witness. There should normally be pre-trial disclosure of expert evidence, in order to save court expense. A party will not normally be allowed to call expert evidence at trial if there has not been disclosure, unless the other side agrees.

In a criminal trial, the prosecution must inform the defence of the name and address of any person who has made a statement related to the prosecution but is not being called as a witness; of the existence of any previous witness statements which are inconsistent with those that person made at the trial; and of any known previous convictions of prosecution witnesses.

1.10.5 Reform

Litigation, whether civil or criminal, is time consuming and expensive. There is ongoing critical discussion about the need for reform in various contexts. The following are among the proposals which might bring about change for health and safety cases:

- The Woolf Report on Access to Justice, which advocates greater judicial (rather than lawyer) control before trial including that the calling of expert evidence should be subject to the complete control of the court.
- The encouragement of alternative dispute resolution (negotiation and arbitration) with civil disputes.
- Proposals to restrict and target legal aid.
- The statutory power[13] for conditional fee agreements linked to a successful outcome, including for personal injury cases.
- Law Commission proposals[14] for the introduction of a special offence of corporate killing where a company's management failure in causing a death fell far below what could be reasonably expected.
- Law Commission proposals[15] for punitive damages to be allowed where employers show 'a blatant disregard of the health and safety of their workforce'.

1.11 Identity of court personnel

1.11.1 The English system

Court personnel include the bench, that is judges or magistrates; counsel for either side (see paragraph 1.8.4); and the court usher appointed to keep silence and order in court, and to attend upon the judge. All judges are appointed by the Crown, and the appointment is salaried and pensionable.

In the Magistrates' Court there are 2–7 Justices of the Peace; or, in London and some large cities, possibly a stipendiary magistrate. Justices of the Peace are lay persons appointed by the Lord Chancellor on behalf of the Queen. The office dates back to the thirteenth century, but is now mainly regulated by the Justices of the Peace Act 1979. Justices sit part-time. They are not paid, but are reimbursed for financial expenses incurred from the office. A stipendiary magistrate is appointed by the Lord Chancellor, and is a qualified solicitor or barrister of at least seven years' standing. The office is salaried and full-time.

A Clerk to the Justices advises justices on questions of law, procedure and evidence; but should not be involved in the magistrates' function of trying the case. Legislation specifies the qualifications for justices' clerks.

Officiating in the county court is a Circuit judge; or a District judge for small claims and interlocutory (pre-trial) matters. A Circuit judge may also sit in the Crown Court. As a result of the Courts and Legal Services Act 1990, eligibility for appointment to the bench is based on having sufficient years of right of audience (qualification) in the courts. A Circuit judge must have 10 years' county court or Crown Court qualification, or be a Recorder, or have held other specified appointments. A District judge requires a 7 year general qualification (i.e. right of audience in any court).

First instance cases in the Crown Court are tried before a judge (to decide on matters of law); and a lay jury (for matters of fact). The Crown Court has three kinds of judge according to the gravity of the offence: a High Court judge, a Circuit judge or a Recorder. A High Court judge (necessary for a serious case) will be a Circuit judge with at least two years' experience, or have a 10 year High Court qualification. A Recorder is part-time, with a 10 year county court or High Court qualification. For appeals to the Crown Court, there will be no jury, but possibly the judge will sit with 2–4 justices.

For the Court of Appeal, normally three judges sit. They are called Lord Justices of Appeal. Appointments are normally made from High Court judges. An alternative prerequisite is 10 years' High Court qualification. High Court judges may also be asked to assist in the Court of Appeal. The Master of the Rolls is president of the Civil Division of the Court of Appeal. The Lord Chief Justice presides in the Criminal Division.

The Appellate Committee of the House of Lords as the final court of appeal sits with at least three 'Law Lords'. The Law Lords include the Lord Chancellor, the Lords of Appeal in Ordinary (who must have held high judicial office for two years or have 15 years' Supreme Court (see para. 1.8.5) qualification, and Peers of Parliament who hold or have held high judicial office.

The head of the judiciary and president of the House of Lords is the Lord Chancellor. He is also a government minister, and the Speaker of the House of Lords. He is exceptional in combining judicial, executive and legislative functions.

The Attorney General is the principal law officer of the Crown. He is usually an MP and answers questions on legal matters in the House of Commons. He may appear in court in cases of exceptional public interest.

His consent is required to bring certain criminal actions, for example in respect of offences against public order. The Solicitor General is immediately subordinate to the Attorney General.

The Director of Public Prosecutions must have a 10 year general qualification. He undertakes duties in accordance with the directions of the Attorney General. He will prosecute cases of murder and crimes amounting to an interference with justice.

1.11.2 Legal personnel in Scotland

In Scotland the Lord Advocate is the chief law officer of the Crown and has ultimate responsibility for prosecutions. He and the Secretary of State for Scotland undertake the duties which in England and Wales are the responsibility of the Home Secretary, the Lord Chancellor and the Attorney General. The Lord Advocate is assisted by the Solicitor General.

Judicial appointment, to the Supreme Court and the Sheriff Court, is by royal authority on the recommendation of the Secretary of State. Judges in the District Courts are lay justices of the peace, apart from some stipendiary magistrates in Glasgow.

The two branches of the legal profession are solicitors and advocates. As in England, advocates no longer have exclusive rights of audience in the higher courts. Traditionally a Scottish solicitor is more a manager of his client's affairs than in England.

1.11.3 Legal personnel in Northern Ireland

The Lord Chancellor, and the English Attorney General and Solicitor General act also for Northern Ireland. The Director of Public Prosecutions is appointed by the Attorney General, and has 10 years' legal practice in Northern Ireland. His chief function is responsibility for prosecutions in serious cases (compare the Crown Prosecution Service in England, and the Lord Advocate and procurators-fiscal in Scotland).

As in England, appointment to the bench and advocacy in the superior courts is at present restricted to barristers. A major difference between the legal system of Northern Ireland and England is the appointment of resident magistrates (RM). They are full-time and legally qualified, with responsibility for minor criminal offences, committal proceedings, and some civil matters. The powers of lay Justices of the Peace in Northern Ireland are limited in comparison with JPs in England and Wales.

1.12 Industrial Tribunals

Industrial Tribunals were set up in 1964 to deal with matters arising under the Industrial Training Act of that year. Now they have statutory jurisdiction in a range of employment matters, such as unfair dismissal, redundancy payments, equal pay and sex and race discrimination. The Secretary of State may by order confer jurisdiction on Industrial Tribunals in respect of claims for breach of contract of employment. Such jurisdiction does not include a claim in respect of personal injuries[16]. In the context of HSW they hear appeals against prohibition and improvement notices, and applications by statutory safety representatives about payment for time off for training.

The burden of proof is on the inspector to satisfy the Tribunal that the requirement for a notice is fulfilled: *Readmans Ltd* v. *Leeds City Council*[17] (a prohibition notice under s. 3). The High Court held that the notice alleged a breach of a criminal duty and it was for the council who had issued the notice to establish the existence of the risk of serious personal injury not for the appellant to have the burden of proving that there was no such risk. The burden of proof is then on an appellant who wishes to show that it was not, for example, practicable or reasonably practicable (according to legislation) to carry out certain measures. This must be proved on the balance of probabilities.

Tribunals sit locally and consist of a legally qualified chairman plus a representative from each side of industry. Proceedings begin with an originating notice of application in which the applicant sets out the name and address of both parties and the facts of the claim. The application must be made within the prescribed time limit. This varies. It is 21 days with enforcement notice; three months for unfair dismissal and paid time off for union duties; six months for redundancy applications.

Proceedings are on oath, but they are more informal than in the courts and the strict rules of evidence are not followed. Legal aid is not available for representation. A friend or union official may represent (which is not possible in the courts). Costs are rarely awarded. Like the courts, Tribunal proceedings are open to the public, and visits are the best way to understand their working.

An appeal is possible from an Industrial Tribunal decision, but only on a matter of law. In respect of enforcement notices it is to the High Court in England; and to the Court of Session in Scotland. In respect of other matters it is to the Employment Appeal Tribunal except in N. Ireland.

The Employment Appeal Tribunal is a superior court associated with the High Court. It sits with a judge and 2–4 lay members, and all have equal voice. Parties may be represented by any person they wish, and legal aid is available. Further appeal is to the Court of Appeal (in Scotland to the Inner House of the Court of Session). In N. Ireland there is no Employment Appeal Tribunal but an Industrial Tribunal's decision may be challenged by review by the Tribunal itself, by judicial review by the High Court, or by way of case stated to the Court of Appeal.

1.13 European Community Courts (ECJ)

1.13.1 The Court of Justice of the European communities

The European Court is the supreme authority on Community law. Its function is to 'ensure that in the interpretation and application of the EEC

Treaty the law is observed' (art. 164). The EEC Treaty and the Single European Act 1986, are concerned with matters such as freedom of competition between Member States; and aspects of social law, including health and safety at work.

The Treaty of European Union 1991 (the Maastricht Treaty) re-emphasised these Community aims and added further goals of economic and monetary union. In *R v. Secretary of State for Transport v. Factortame Ltd*[18], the ECJ directed the House of Lords that any provision of a national legal system which might impair the effectiveness of EU law is incompatible with the requirements of EU law. UK regulations made under the Merchant Shipping Act 1988, to prevent 'quota hopping' by Spanish fishermen, were struck down as being contrary to EU law. In *Factortame No. 5*[19], the Court of Appeal held that the breaches of Community Law were sufficiently serious to give rise to liability for damages to individuals[20].

The European Court has two types of jurisdiction, direct actions, and reference for preliminary rulings.

Direct actions may be:

- against a Member State for failing to fulfil its obligations under Community law and be brought by the Commission or by another Member State;
- against a Community institution, for annulment of some action, or for failure to act (judicial review);
- against the Community for damages for injury by its institutions or servants;
- against a Community institution brought by one of its staff.

References for preliminary rulings are requests by national courts for interpretation of a Community provision. Article 177 provides that any court or tribunal may ask the European Court for a ruling; but only the final court of appeal (the House of Lords in the UK) must ask for a ruling if a party requests it. In the English case of *Bulmer v. Bollinger*[21] the Court of Appeal held that the High Court and the Court of Appeal may interpret Community law.

The European Court is based at Luxembourg. There are 13 judges (to include one from each Member State), assisted by six Advocates General. The function of an Advocate General is to assist the Court by presenting submissions, in which he analyses the relevant issues and makes relevant recommendations for the use of the Court. The judgement itself is a single decision, thus an odd number of judges is required. With the increase in workload, there is a facility for the Court to sit in subdivisions called Chambers. Cases brought by a Member State or by a Community institution must still be heard by the full court. Although the Court seeks to have consistency in its findings, precedents are persuasive rather than binding on itself. Decisions are binding on the particular Member State.

Referrals to the Court of Justice are requests to it to rule on the interpretation or applicability of particular parts of Community law. Where the Court of Justice makes a decision, it not only settles the particular matter at issue but also spells out the construction to be placed on disputed passages of Community legislation, thereby giving clarification and guidance as to its implementation.

It keeps under review the legality of acts adopted by the Council and the Commission and also can be invited to give its opinion on an agreement which the Community proposes to undertake with a third country, such opinions become binding on the Community.

Through its judgments and interpretations, the Court of Justice is helping to create a body of Community law that applies to all Community institutions, Member States, national governments and private citizens. Judgements of the European Court of Justice take primacy over those of national courts on the matters referred to it.

Although appointed by the Member States, the Court of Justice is not answerable to any Member State or to any other EC institution. The independence of the judges is guaranteed.

Under the Single European Act 1986, the Council of Ministers has the power to create a new Court of First Instance. This Court was established by Council decision in 1988 and became effective in September 1989. It has 12 members, appointed by common accord of the Member States. Members may also be asked to perform the task of an Advocate General. It may sit with three or five judges.

The jurisdiction is:

- disputes between the Community and its staff;
- applications for judicial review against the Council or Commission;
- applications for judicial review in some matters against the European Coal and Steel Community.

There is also a Court of Auditors, which supervises the implementation of the budget.

At the beginning of this section, the *Factortame* litigation illustrated the interaction of ECJ decisions with UK courts. Another illustration of the effect of an ECJ decision on national law comes from the UK challenge to the *Working Time Directive*. The ECJ rejected the UK argument that the legal basis of the directive was defective, and also considered that the directive did not breach the principle of *subsidiarity* (the aims could not be achieved by Member States alone), nor *proportionality* (the requirements were not excessive). The directive is now being implemented by national UK regulations.

1.13.2 The European Court of Human Rights

This Court should not be confused with the Court of Justice of the European Communities. The Court of Human Rights sits at Strasbourg. Its function is to interpret the European Convention for the Protection of Human Rights and Fundamental Freedoms, drawn up by the Council of Europe in 1950. The Council of Europe comprises 23 Western European states. It is active on social and cultural fronts rather than economic. The United Kingdom ratified the Convention in

1951, so that it is binding on the UK internationally. However, UK legislation has not yet incorporated the Convention. The articles of the Convention provide for matters such as the right not to be subjected to inhuman or degrading treatment, the right to freedom of peaceful assembly, the right to respect for family life, home and correspondence.

An example of a decision directed to the UK was the *'Sunday Times thalidomide case'* in 1981. A drug prescribed for pregnant women caused severe abnormalities in the children. The manufacturers sought an injunction to prevent the *Sunday Times* publishing an article about the drug. The Court of Human Rights ruled that the House of Lords' confirmation of an injunction was a violation of the right of freedom of expression[22].

1.14 Sources of English law

The two main sources of UK law are legislation, and legal principles developed by court decisions (common or case law).

English common law, based on custom and evolving since the eleventh century, developed indigenous concepts, and unlike most European countries was little influenced by Roman law. In Scotland Roman law was an important influence from the sixteenth to the eighteenth century, particularly on the law of obligations, which includes contract and delict. In Ireland, before the seventeenth century, Brehan law (of early Irish jurists) or English common law predominated according to political control at the time. Since the seventeenth century the law in Ireland and England developed along similar lines in general, with some exceptions such as marriage and divorce. English common law concepts were applied in former British territories. Today most of the United States, Canada (other than Quebec), Australia, New Zealand, India and some African countries remain and are called common law countries.

England, Scotland and N. Ireland do not have codified legal systems. Nearly all of our law of contract and much of the law of tort or delict is case law. This will gradually change with the production and implementation of Law Commission reports.

As with most subjects, law has specific terminology. The historic development of our law is illustrated by the Latin, old French and old English phrases which are sometimes used. This chapter contains some Latin words, for example, *obiter dicta* and *ratio decidendi* (section 1.9); and some coming from the French, such as tort and plaintiff (sections 1.5, 1.6). The most straightforward rule for legal Latin or French is to pronounce words as though they were English. Other words and phrases met with have a particular legal meaning, such as damages, contract of employment, relevant statutory provision; and abbreviations such as JP or *v.* (as in Donoghue *v.* Stevenson). There are a number of law dictionaries to explain or to translate words and these are listed at the end of this chapter.

1.15 Legislation

1.15.1 Acts of Parliament and delegated legislation

Since the eighteenth century increasing use has been made of legislation. Legislation comprises Acts of Parliament and delegated legislation made by subordinate bodies given authority by Act of Parliament. Examples of delegated legislation are ministerial orders and regulations (Statutory Instruments), local authority byelaws and court rules of procedure. All legislation is printed and published by The Stationery Office Ltd. Often, but not always, delegated legislation requires the approval of Parliament, for example by negative resolution (that is by not receiving a negative vote of either House); or, more rarely, by affirmative resolution (that is by requiring a positive vote of 'yes').

HSW and its associated regulations is an example of how extensive subordinate legislation may be. HSW is an enabling Act. Section 15, schedule 3 and s. 80 give very wide powers to the Secretary of State to make regulations. The regulations are subject to negative resolution (s. 82). They may be made to give effect to proposals of the Health and Safety Commission (in N. Ireland the Health and Safety Agency); or independently of such proposals, but following consultation with the Commission and such other bodies as appear appropriate (s. 50). The Commission may also issue Approved Codes of Practice (s. 16 HSW) for practical guidance. Such codes are not legislation and s. 17 confirms that failure to observe such codes cannot of itself ground legal proceedings. However, failure to comply is admissible evidence and will be proof of failure to comply with a legislative provision to which the code relates unless the court is satisfied that there is compliance in some other way.

Delegated legislation is suitable for detailed technical matters. By avoiding the formality required for an Act of Parliament the legislation can be adapted, and speedily (for example, the maximum unfair dismissal payment may be increased quickly by an Order).

Long drawn out consultation may slow down any legislation. In 1955 the decision in a famous case of *John Summers & Sons Ltd* v. *Frost*[23] virtually meant that an abrasive wheel was used illegally unless every part of that dangerous machinery was fenced. Regulations were required to allow its legal use. There were drafts and consultations, but it was 1971 before the Abrasive Wheels Regulations came into operation[24].

During its passage through Parliament and before it receives the Royal Assent an intended Act is called a *bill*. Most government bills start in the House of Commons, but non-controversial ones may start in the House of Lords. Ordinary public bills such as that for HSW go through the following process. The bill is introduced and has a formal first reading. At the second reading there is discussion on the general principles and purpose of the bill. It then goes to committee. After detailed consideration the committee reports the bill to the House, which considers any amendments. The House may make further amendments and return the bill to committee for further consideration. After the report stage the bill is read for the third time. At third reading in the Commons only verbal alterations may be made.

The bill now goes through similar stages in the other House. If the second House amends the bill it is returned to the first House for consideration. If the Lords reject a bill for two sessions it may receive the Royal Assent without the Lords' agreement. Practically, the Lords can delay a bill for a maximum of one year.

After being passed by both Houses the bill receives the Royal Assent, which conventionally is always granted, and thus becomes an Act. A statute will normally provide at the end whether it is to apply in Scotland and N. Ireland as well as in England and Wales. Subsequent legislation may apply provisions to Scotland or N. Ireland, for example the Health and Safety at Work (NI) Order 1978.

Parliament has supreme authority. It may enact any measure, other than binding future Parliaments. It is not answerable to the judiciary.

The United Kingdom is now part of the European Community (EU) and subject to the Community's regulations and directives (see paragraphs 1.13.1 and 1.15.4). These require Member States to implement agreed standards on, among other concerns, safety and health at work and the environment.

The ultimate sovereignty of the UK Parliament is theoretically retained in that Parliament could repudiate agreement to EU membership[25]. Also, since the Single European Act there has been increased emphasis on *subsidiarity*. This is the principle that decisions should be taken at the most suitable level down the hierarchy of power, that is at national rather than EC level where appropriate.

1.15.2 Statutory interpretation

Inevitably some legislation has to be interpreted by the courts, to clarify uncertainties, for example, and substantial case law may attach to a statute. Judicial consideration of the effect of legislation for the fencing of dangerous machinery is an example of this (see sections 1.16.3 and 1.18.1).

Statutes normally contain an interpretation section. There is also the Interpretation Act 1978 which provides, for example, that unless the contrary is stated, then male includes female, the singular includes the plural, writing includes printing, photography and other modes of representing or reproducing words in visible form. In modern legislation, the detail is often relegated to Schedules at the end of the Act.

Parliamentary discussions are reported verbatim in *Hansard*. In 1992 the House of Lords decided that if there is an ambiguity, a minister's clear explanation to Parliament, as published in *Hansard*, may be used to interpret a statute[26].

As a result of Article 5 of the EEC Treaty 1957, which requires Member States to 'take all appropriate measures to ensure fulfilment of the obligations arising out of the treaty', UK courts give a *purposive* interpretation where the purpose of UK legislation is to give meaning to a directive. An example is *Pickstone v. Freeman plc*[27]. The House of Lords interpreted regulations amending the Equal Pay Act against their literal meaning to allow a female warehouse operative to use as a comparison a man doing a different job of equal value.

1.15.3 White Papers and Green Papers

Proposed legislation may be preceded by documents presented by the government to Parliament for consideration. A Green Paper is a discussion document. A White Paper contains policy statements and explanations for proposed legislation. Such papers are published as Command Papers.

On a narrower basis the Government also consults with outside interests when drafting legislation, bodies such as the CBI and TUC on industrial and economic matters. Legislation may require such consultation, for example s. 50 HSW.

1.15.4 European Union (EU) legislation

Originally known as the European Economic Community (EEC) and then as the European Community (EC) it is now usual to refer to the Community as the European Union (EU). The primary legislation is the Treaty of Rome 1957 which established the Community and was incorporated into UK law by the European Communities Act 1972; the Single European Act 1985 was incorporated into UK law in 1986 and the Treaty of European Union 1991 by the European Communities Act 1992.

Secondary community legislation takes three forms: Regulations which are binding on Member States, Directives which require national implementation (see section 1.15.1 and *Figure 1.10*) and Decisions of the Council or Commission. Such a decision is specific rather than general. Its main use is if a State asks permission to depart from the EEC Treaty, for example in respect of competition policy.

Legislation is usually initiated by the European Commission and, after statutory consultation is adopted by the Council of the European Community (CEC). The administration of the EU is in the hands of the Commission which has 17 members, one from each Member State but with the larger Member States having two. The supreme body of the EU is the Council of Ministers with one member from each State but with weighted voting rights according to size. The Council receives proposals from the Commission and consults with the Economic and Social Committee (EcoSoC) and the European Parliament with the aim of reaching a common position on the proposal. When the Council adopts a proposal it places obligations on Member States to incorporate its requirements into national laws within a stated time scale. Adopted legislation is published in the Official Journal of the European Communities.

In outline, the procedure for secondary legislation is that the Commission proposes and consults; the European Parliament considers

1.32 Safety Law

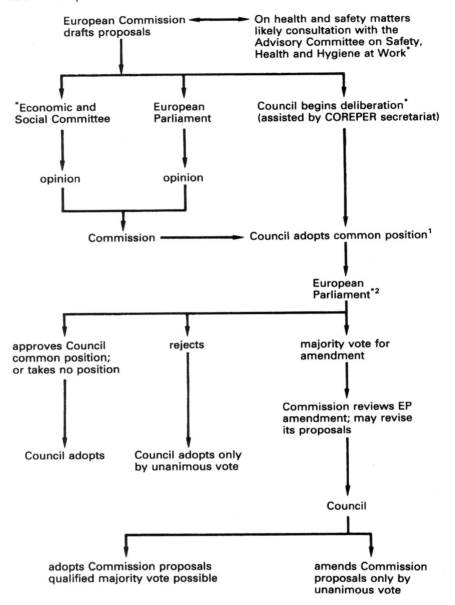

Figure 1.9 EC co-operation procedure for adoption of directives

Community adopts Directive
(member States must implement within time limit)

1. HSE prepare proposals
2. Limited consultations with e.g. CBI, TUC, professional bodies
3. Draft regulations to HSC
4. HSC issues consultative document for public comment (consultation period of some 4 months)
5. Comments co-ordinated by HSE
6. Finalised proposals, taking account of public comment
7. Final proposals submitted to HSC
8. HSC submit proposals to Secretary of State
9. Proposals placed before parliament; negative resolution
10. Put on Statute Book and becomes UK law
11. Effective on date announced

Figure 1.10 Stages of internal UK procedure for implementing a directive

and may propose amendments; the Council adopts; and Member States implement. *Figure 1.9* illustrates this in more detail. *Figure 1.10* shows internal UK procedure for incorporating a directive into UK law.

The function of the European Parliament is advisory and supervisory rather than legislative and much of its work is done in committees. Parliamentary influence has increased following the introduction of direct elections in 1979 and the Single European Act in 1986.

The Single European Act introduced, for certain matters, a 'co-operative' procedure between Council, Parliament and the Commission which allows Parliament a second chance to comment and suggest changes to proposals once a common position has been adopted by Council. If the Commission accepts these changes then Council can adopt the proposal by qualified (weighted) majority vote. If the Commission and Parliament cannot reconcile their opinions then Council can only adopt the proposal by unanimous vote. This new procedure applies, among other matters, to proposals concerning the health and safety of workers. The introduction of qualified majority voting was the stimulus for a great increase in EU health and safety directives from 1989 onwards.

The restricted power of the elected body, the European Parliament, causes concern that the EU lacks democratic accountability. The Maastricht Treaty (Treaty on European Unity 1991) includes greater power for the European Parliament, allowing it to reject certain proposals, including those on health and consumer affairs (Article 189b). Also, Article 138b provides that the European Parliament may, acting by majority vote of its members, request the Commission to prepare appropriate proposals on any matters on which it considers that a Community act is required for the purposes of implementing the Treaty.

The European Agency for Safety and Health at Work was officially inaugurated in 1997[28]. Its functions include assessment of the impact of health and safety legislation on small and medium enterprises and the establishment of a network to share information within the EU and more widely.

1.15.5 Application of EU legislation to an individual

The Treaty and Community legislation must be recognised in the Member States, but an individual can only enforce it, if at all, in the national courts; and only if it has 'direct effect' for that individual. Community legislation takes two main forms, regulations and directives (see also paragraph 1.15.4). A regulation is a law in the Member States to which it is directed; it is said to be 'directly applicable' to that State. According to its content a Community regulation may impose obligations and confer rights on individuals enforceable in the national courts; it is then said to have 'direct effect'. A directive must be enacted by the Member State, and then, according to how it is enacted, may give enforcement rights to individuals in the national courts. Sometimes a directive, even before implementation by the Member State, may have 'direct effect' for an individual to rely on it against the State. This could be so if the date of implementation had passed and the existing law of the Member State contravenes the directive[29]. The directive must be sufficiently clear, precise and unconditional.

Any such direct effect of a directive does not give rise to obligations between individuals. However, in *Marshall* v. *Southampton and South West Hampshire Area Health Authority (Teaching)*[30], Mrs Marshall successfully challenged the health authority's compulsory retiring age of 65 for men and 60 for women as being discriminatory. An individual may not enforce such a decision against a private employer but can against a government body[31]. See also the repercussions of the *Factortame* case outlined in section 1.13.1. However, the European Court of Justice has required national courts to *interpret* national legislation to be consistent with directives.

1.16 Safety legislation before the Health and Safety at Work etc. Act

1.16.1 Factories

Early factory legislation, in the nineteenth century, concerned the textile and allied industry. It was directed towards the protection of young persons and women and was motivated by concern for moral welfare and sanitation as much as for safety. Between 1875 and 1937 there were attempts to unify the increasing but fragmented legislation, but subsequent inadequacies resulted in patchwork amendments. The Factories Act 1937 was intended as a coordinating measure. It brought together health, safety and welfare in all factories: and introduced some new requirements such as those for floors, passages and stairs, and for safe access.

But regulations made under previous legislation continued in force as though made under the 1937 Act. This practice was repeated by the Factories Act 1961 so that some of the provisions and standards were outdated. The HSW and consequent regulations, including those implementing EC directives, have replaced much of the Factories Act and associated legislation.

Similarly, HSW regulations have superseded or augmented other workplace-specific provisions, such as for offices, agriculture, mines and quarries.

1.16.2 Offices

In 1949 the Gower Committee report made recommendations about the health, welfare and safety of employed persons outside the protections of existing legislation. In 1960 an Offices Act was passed. Before it became operative, however, it was repealed and replaced by the Offices, Shops and Railway Premises Act 1963. This adopted much of the structural content of the Factories Act 1961 but not the regulations, which apply only to factories.

1.16.3 Mines, quarries etc

The law relating to safety and management in mines and quarries was examined in the 1950s and the principal Act is now the Mines and Quarries Act 1954. HSW regulations are more likely to augment and update rather than absorb rules for this very particular work environment. There is wide power to make regulations. Other Acts refer to work practices in agriculture, aviation and shipping.

1.17 Safety legislation today

1.17.1 Health and Safety at Work etc. Act 1974

In 1970 the Robens Committee was set up to review the provision made for the safety and health of persons in the course of their employment. At that time safety requirements were contained in a variety of enactments (as the list of relevant statutory provisions in schedule 1 of HSW indicates). An estimated five million employees had no statutory protection. Protection was uneven. Administration was diverse and enforcement powers were considered inadequate. The wording and intent of the legislation were not directed towards personal involvement of the worker; and in parts it was obsolete.

HSW corrects many of these defects. General principles are enacted, to be supplemented by regulations. The provisions apply to employments generally to protect persons at work and those at risk from work activities.

The Act was intended to be wide to facilitate changing circumstances. Examples of development are the sanctions for non-compliance; and the use of the extensive powers to make regulations under s. 15 and Schedule 3.

Magistrates may now impose a fine up to £20 000 for breach of ss. 2–6 HSW or for a breach of an improvement or prohibition notice or a court remedy order. In addition, magistrates may imprison individuals for up to six months for breach of an improvement or prohibition notice or court remedy order[32].

Sections 2–6 were selected because they contain the main health and safety duties of those responsible for workplace safety. It was considered that a company charged with breach of one of these sections is probably responsible for a systematic failure to meet these general duties and is putting its employees and possibly others at risk. Failure to comply with a notice indicates a deliberate flouting of health and safety law.

The maximum magistrates' fine for other offences is £5000[33].

The Crown Court has used powers under s. 2(1) of the Company Directors Disqualification Act 1986 to disqualify a director for two years. The Act allows the court to make a disqualification order against a person convicted of an indictable offence connected with the *management* of a company. The accused's company was fined under s. 33 HSW for breach of a prohibition notice; and the accused under s. 37 HSW because the company's offence was committed with his consent, connivance or attributable to his neglect[34].

1.17.2 EU influence

The Single European Act 1986, with the objective of a single market by 1 January 1993, has had a dynamic effect on the introduction of health and safety legislation. The implementation of effective common health and safety standards is considered conducive to attaining a 'level playing field' for employers across the Community; and to the participation of the workforce in the intended resulting economic benefits.

Article 118A (introduced by the 1986 Act) provides that Member States shall 'pay particular attention to *encouraging improvements, especially in the working environment, as regards the health and safety of workers*, and shall set as their objective the harmonisation of conditions in this area, while maintaining the improvements made'.

A change in EU approach has been the use of Framework and related 'Daughter' Directives. The Framework Directive on the introduction of measures to encourage improvements in the safety and health of workers at work, with five daughter directives is an example[35]. The directive has been implemented in the UK as the Management of Health and Safety at Work Regulations 1992 (MHSW). The core of these regulations is the duty to assess the risks to health and safety to employees and anyone who may be affected by the work activity, and to follow through with appropriate measures of planning, care and information.

Implementation has been possible under HSW. Section 1(2) provides for the progressive replacement of existing legislation by a system of regulations and approved codes of practice 'designed to **maintain or improve** the standards of health, safety or welfare established by or under those enactments'.

There are a number of further directives and draft directives relevant to health and safety. The HSC will not negotiate all the implementation. For example, a number of the 'Technical Harmonisation and Standards' directives are co-ordinated by the DTI. The HSC and HSE are involved in consultation.

1.17.3 Standards of duty

In criminal and in civil actions the person alleging a breach has the burden of proof, i.e. must prove the wrongdoing. This burden is more easily discharged if an offence is 'absolute' which means that proof of the commission of the act is enough for liability. In criminal law the prosecution must normally prove guilty intent (*mens rea*) in addition to the guilty act (*actus reus*). If exceptionally, guilty intent need not be proved, the crime is described as absolute. In that sense, the Health and Safety at Work Act (HSW) imposes absolute duties. This was emphasised in *R. v. British Steel plc*[36] where the Court of Appeal held that it was not necessary to find a company's 'directing mind' (its senior management) at fault in order to prove the company's liability.

Although corporate liability is absolute in the above sense, most of the general duties of HSW (and some of the duties of the regulations[37]) are qualified by the defence that steps must be 'reasonably practicable'. This has been interpreted to mean that the risk should be balanced against the 'cost' of the measures necessary to avert the risk (whether in money, time or trouble) to see if there is gross disproportion[38].

Other duties are qualified by 'practicable'. This is a stricter duty than reasonably practicable and has been interpreted to mean not as arduous as physically possible. A measure is practicable if it is possible in the light of current knowledge and invention[39].

The description 'strict' liability is sometimes used in the same sense as 'absolute' liability (to apply to criminal offences where there is no requirement of *mens rea*). However, 'absolute' and 'strict' are sometimes differentiated so that absolute is used in a narrow sense to mean that there is no defence if the act is proved, although there may be a defence in strict liability. Section 9 HSW, the duty not to charge an employee for things provided because of a specific statutory requirement, has been suggested as a rare example of 'absolute' in the narrow sense. In contrast, an employer's duty to undertake a suitable and sufficient risk assessment of his/her undertaking for employees and others[40] is strict. However, the approved code of practice[41] suggests risk 'reflects both the likelihood that harm will occur and its severity'. That will affect whether the assessment is suitable and sufficient. Another example is the strict duty on the employer in reg. 11 of the Provision and Use of Work Equipment Regulations 1992 (PUWER) to prevent access to any *dangerous* part of machinery. 'Dangerous' is not defined in those regulations but it was interpreted in the now repealed s. 14 of FA as being 'a reasonably foreseeable cause of injury to anybody acting in a way in which a human being may reasonably be expected to act in circumstances which may reasonably be expected to occur'[23]. The duty is strict but subject to a 'reasonable foreseeability' test.

In civil law involving personal accidents (the law of tort) strict liability is unusual. A plaintiff must normally prove fault, in the form of negligent conduct of the defendant, which is assessed objectively.

Some apparently strict duties of EU health and safety directives have been transposed into UK legislation as being reasonably practicable. The HSE has explained that this is to avoid conflict of two absolute duties. For example Article 3 of the EU manual handling of loads directive[42] requires the employer to use appropriate means to avoid manual handling and to take steps to control manual handling that does take place. European law is accustomed to deal with such conflicts with the doctrine of *proportionality*, that is balancing consequences to see whether an absolute ban is disproportionate to a goal which could be achieved by less restrictive means.

1.18 Principles developed by the courts

1.18.1 Case law interpretation

Case law interpretation has had an adverse effect on some safety legislation. A notorious example is the fencing requirements for dangerous machinery (then s. 14 FA), as illustrated by, for example, *Close* v. *Steel Company of Wales*[43]. With reluctance judges interpreted the statute so that s. 14 could not be used where parts of the machine or of the material being worked on have been ejected at a workman. This interpretation has now been remedied by reg. 12(3) of PUWER.

Such interpretations affect the scope of legislation, and of civil action for breach of statutory duty. Breach of statutory duty and the tort of negligence are the two most frequent grounds for civil claims following accidents at work. As identified in section 1.6, an employee's contract of employment is important for the duties owed by the employer.

1.18.2 Tort of negligence

Negligence is a relatively modern tort, but today it is probably the most important in number of cases and for the amount of damages which may be awarded for serious injury.

The tort consists of a breach by the defendant of a legal duty to take care not to damage the plaintiff or his property and consequent damage from that breach. From early times the common law has placed on the employer duties towards his employees. In 1932, Lord Atkin, in the leading case of *McAlister (or Donoghue) v. Stevenson*[44] suggested a general test for when a duty is owed. It is owed to persons whom one ought reasonably to have in mind as being affected by the particular behaviour. In 1963 the persuasive precedent of *Hedley Byrne* v. *Heller & Partners*[3] extended the duty to include financial loss resulting from some careless statements.

Since 1988[45] the potentially wide scope of the duty of care has been narrowed so that there are now four indicators: foresight of damage, proximity of the defendant to the plaintiff, policy and whether it is just and reasonable to impose a duty. A court will not necessarily refer to them all in the same case, but will look at the particular relationship. An important one is that of employer and employee. The duty of care owed to an employee is an implied term of the contract of employment (see section 1.18.4). In respect of premises, the common law duty of care owed by the occupier is now statutory (see section 1.18.5).

Examples of health concerns, developed in the civil tort of negligence and which are receiving increasing attention in the courts and by the HSE, are workplace stress; repetitive strain injury (RSI) and (WRULD); and vibration white finger (VWF)[46].

1.18.3 Tort of breach of statutory duty

When a statutory duty is broken there is liability for any penalty stipulated in the statute. In addition a person suffering damage from the breach may sometimes bring a civil action in tort to obtain compensation. Sometimes the Act specifies this (for example, the Consumer Protection Act 1987). Sometimes the Act is silent but the courts allow the action, as happened with FA and related regulations; or the Act is silent but the courts deny a civil action. This happened with the Food and Drugs Act 1955 (which has now been consolidated with other enactments relating to food into the Food Safety Act 1990) when it was decided that the statute was not intended to add to a buyer's civil remedies for breach of contract or of negligence.

Section 47 of HSW provides that breach of the Act will not give rise to a civil action, but breach of any regulation made under the Act is actionable, unless the regulations say otherwise. So far the only regulations to provide otherwise are the MHSW[47].

Negligence and breach of statutory duty are two different torts, but both may be relevant following an incident. Bertha, injured at work because of an obstruction of the factory floor, might allege negligence plus breach of reg. 12 of the Workplace (Health, Safety and Welfare) Regulations 1992 (WHSW), and possibly succeed in both torts. She would not recover double damages because the remedy is compensation for the actual loss suffered.

1.18.4 The contract of employment

Implied terms of the contract of employment include the common law requirements that *employers* take reasonable care of the safety of employees and do not undermine the trust and confidence of the employee. The former duty has three connected requirements – the provision of competent fellow workers, safe premises, plant and equipment and a safe system of work. An employer cannot delegate this duty to another[48].

This implied contractual duty is the basis of the legal duty of care to an employee in the tort of negligence. The concept has extensive implications. For example, the Court of Appeal has said that a contract requiring long hours of work from a junior doctor is subject to the implied duty of care not to harm an employee[49]. In a successful constructive dismissal claim based on passive smoking[50], the Employment Appeal Tribunal (finding guidance from s. 2(2)(e) HSW) suggested that the implied contractual duty in any employment contract encompassed an implied term that the employer will provide and maintain, so far as is reasonably practicable, a working environment that is reasonably suitable for the performance of an employee's duties.

1.18.5 Duty to third parties on site

Third parties may be on premises with the occupier's *express* consent. Examples include customers, independent contractors and their employees, business associates or non-executive directors. Others such as an inspector or the postman may be on the premises with the occupier's *implied* consent. There may also be trespassers on the premises without express or implied permission, this category including those exceeding their invited purpose, such as customers entering the stock room, for instance.

The common law duty of care owed to visitors by an occupier in respect of premises is now statutory and was clarified in the *Occupier's Liability Act 1957* which ended the previous (often subtle) distinction between persons invited to enter (called *invitees*) and those allowed to enter (*licensees*), a distinction which previously affected the standard of duty. Under the 1957 Act, both categories are visitors to whom an occupier owes the 'common duty of care' once the relationship of occupier and visitor is established. The duty is to take such care as in all the circumstances of the case is reasonable to see that the visitor will be reasonably safe in using the premises for the purposes for which he is invited or permitted to be there. An example is *Cunningham* v. *Reading Football Club*[51]. Due to the football club's failure to maintain its terraces, football hooligans were able to use lumps of masonry as missiles. A policeman on duty at the club was injured and successfully sued that club.

The 1957 Act makes specific reference to visitors present in the exercise of their calling who may be expected to appreciate and guard against any special risks incidental to that calling, and to child visitors.

The *Occupier's Liability Act 1984* now applies to 'persons other than visitors'. As well as trespassers, this category also includes persons merely exercising a right of way across premises. The 1984 Act provides that there is a duty owed to uninvited entrants if the occupier has reasonable grounds to believe a danger exists on his premises and the consequent risk is one against which, in all the circumstances, he/she may reasonably be expected to provide some protection.

Aside from the duties as occupier, the tort of negligence continues to apply for whoever creates a source of danger. In the criminal context, HSW[52] and the relevant regulations also apply when a contractor is employed, for example.

1.18.6 Defences

There are two general defences to a civil action for the torts of negligence and breach of statutory duty. The defence that the negligent behaviour of the plaintiff contributed to the result allows the court to reduce a damage award proportionately. The defence of consent to the risk (*volenti non fit injuria*) negates liability. Consent is more than knowledge and this defence rarely succeeds against an employee, because employees may feel constrained in how they undertake tasks. Additionally, there may be specific defences to allegations of breach of statutory duty such as the defence of reasonable practicability.

Statute limits the time within which an action may be brought. For personal injuries the time limit is three years from the date of the breach or from the date of knowledge (if later) of the person injured. The plaintiff must prove every element of an allegation, including that the injury (physical or financial) was consequent on the breach. Thus 'no causation' may be a defence[53].

In criminal prosecutions, the absence of any element of an offence will provide a specific defence to a criminal charge. The time limit for a prosecution of a summary offence in a magistrates' court is six months from the date of the offence. (There is no time limit for Crown Court prosecutions). Statute may provide specific defences, for example HSW generally allows 'not reasonably practicable' as a defence. Some of the health and safety regulations (though not HSW) have a 'due diligence' defence, for example the Control of Substances Hazardous to Health Regulations 1994 (COSHH) provide that 'it shall be a defence for any person to prove that he took all reasonable precautions and exercised all due diligence to avoid committing an offence'.

The fact that an accident has occurred and resulted in legal action being taken is unsatisfactory. An award cannot repair an injury; the outcome of an action is uncertain; and the considerable cost and ingenuity expended in the investigation, developing the pleadings and the trial itself, could have been used more positively in trying to avoid such accidents. Such avoidance is an objective of HSW; and of the EC Directives, which are having increasing importance.

Because of the constraints of space, this chapter can be an outline only. Students are recommended to complement the chapter with further reading (see below) and visits to courts and tribunals.

References and endnotes

1. Powers of Criminal Courts Act 1973, Criminal Justice Act 1991, HMSO, London
2. R. *v.* George Maxwell Ltd (1980) 2 All ER 99
3. Hedley Byrne & Co. Ltd *v.* Heller & Partners Ltd (1964) AC 463
4. *Current Law*: a monthly publication from Sweet and Maxwell
5. Operated by Butterworth (Telepublishing) Ltd
6. For example: Roberts Petroleum Ltd *v.* Bernard Kenny Ltd (1983) 1 All ER 564 HL
7. Waugh *v.* British Railways Board (1979) 2 All ER 1169
8. See s. 33(2) HSW for such offences
9. Criminal Justice and Public Order Act 1994. The court may 'draw such inferences as

appear proper from a failure to mention facts relied on in his defence (s. 34), and if the accused does not give evidence or answer questions without good cause (s. 35)'
10. Criminal Evidence Act 1898 section 1.f
11. The Police and Criminal Evidence Act 1984
12. There are special rules about children, the accused and the accused's spouse
13. The Courts and Legal Services Act 1990
14. Law Commission, Consultation Paper No. 135, *Manslaughter* (1994) and No. 237, *Legislating the Criminal Code; Involuntary Manslaughter* (1996)
15. Law Commission No. 247, *Aggravated, exemplary and restitutionary damages*, HC 346
16. Employment Protection Consolidation Act 1978 and Trade Union Reform and Employment Rights Act 1993
17. Readmans Ltd *v*. Leeds City Council (1992) COD 419
18. R. *v*. Secretary of State for Transport *v*. Factortame Ltd C 221/89; (1991) 1 AC 603; (1992) QB 680
19. Factortame Ltd No. 5, Times Law Reports, 28 April 1998
20. Liability for any damage caused to trawler owners and managers refused registration
21. Bulmer *v*. Bollinger (1974) 4 All ER 1226
22. AG *v*. Times Newspapers Ltd (1979) 2 EHRR 245, European Court of Human Rights
23. John Summers & Sons Ltd *v*. Frost (1955) AC 740
24. Much of the Abrasive Wheels Regulations 1970 has been replaced by the Provisions and Use of Work Equipment Regulations 1992 and the Workplace (Health, Safety and Welfare) Regulations 1992
25. McCarthys Ltd *v*. Smith (1979) 3 All ER 325
26. Pepper *v*. Hart (1992) NLJ Vol. 143 p. 17
27. Pickstone *v*. Freeman plc (1989) 1 AC 66
28. European Agency for Safety and Health at Work authorised by European Council Regulation No. 1643/95
29. Van Duyn *v*. Home Office (Case 41/74) (1975) 3 All ER 190
30. Marshall *v*. Southampton and South West Hampshire Area Health Authority (Teaching) (1986) case 152/84 1 CMLR 688; (1986) QB 401
31. Rolls Royce plc *v*. Doughty (1992) ICR 538
32. by s. 4 of the Offshore Safety Act 1992
33. Effective since October 1992 by the Criminal Justices Act 1991
34. Rodney Chapman at Lewes Crown Court on 26.6.92. In 1991 an Employment Minister said in Parliament that the potential scope of s. 2(1) of HSW was understood to be very broad and that 'management' includes the management of health and safety
35. EC Directive No. 89/391/EEC, adopted 12.6.89 with five daughter directives
36. R. *v*. British Steel plc (1995) ICR 587. This was a prosecution under s. 3 HSW following the death of two employees of a subcontractor employed by British Steel to reposition a steel platform. The contractor's procedure was inherently dangerous but the contract provided for the supervision of the work by a British Steel employee
37. For example, the Manual Handling Operations Regulations 1992
38. Edwards *v*. National Coal Board (1949) 1 All ER 743
39. Adsett *v*. K & L Steelfounders and Engineers Ltd (1953) 1 All ER 97; 2 All ER 320
40. Regulation 3 of the Management of Health and Safety at Work Regulations 1992
41. Health and Safety Executive, Legal Series Book No. L21, *Management of health and safety at work. Approved Code of Practice*, HSE Books, Sudbury (1992)
42. Regulation 4 of the Manual Handling Operations Regulations 1992
43. Close *v*. Steel Company of Wales (1962) AC 367
44. Donoghue *v*. Stevenson (1932) AC 562
45. Murphy *v*. Brentwood District Council (1991) AC 398
46. In addition to grounding a civil action, the statutory requirements in various regulations for employers to assess and to have a policy to deal with risks could now be relevant to such situations
47. Regulation 15 of the Management of Health and Safety at Work Regulations 1992
48. Wilsons & Clyde Coal Co. Ltd *v*. English (1938) AC 57, HL
49. Johnstone *v*. Bloomsbury Health Authority (1992) QB 333
50. Waltons & Morse *v*. Dorrington (1997) IRLR 488
51. Cunningham *v*. Reading Football Club (1991) *The Independent*, 20 March 1991
52. ss. 3 and 4 HSW, for example
53. Corn *v*. Wier's Glass Ltd (1960) 2 All ER 300

Further reading

Atiyah, P.S. and Cane, P., *Accidents, Compensation and the Law*, 5th edn, Butterworths, London (1993)
Barnard, D. and Houghton, *The New Civil Court in Action*, Butterworths, London (1993)
Barnard, D. O'Cuin and Stockdale, *The Criminal Court in Action*, 4th edn, Butterworths, London (1996)
Barrett, B. and Howells, R., *Occupational Health and Safety Law*, Pitman Publishing, London (1997)
Cope, C. and Thomas, P., *How to Use a Law Library*, 3rd edn, Sweet & Maxwell, London (1996)
Dewis and Stranks, *Tolley's Health and Safety at Work Handbook*, Tolley, London (1998)
Dickson, B., *The Legal System of Northern Ireland*, 3rd edn, SLS Publications (NI), Belfast (1993)
Encyclopaedia of Health and Safety at Work, Sweet and Maxwell, London (loose-leaf)
Hutchins, E.L. and Harrison, A., *History of Factory Legislation*, F. Cass, London (1966)
James, P., *Introduction to English Law*, 13th edn, Butterworths, London (1996)
Munkman, J., *Employer's Liability at Common Law*, 12th edn, Butterworths, London (1996)
Selwyn, N., *The Law of Employment*, 10th edn. Butterworths, London (1998)
Smith, K.C. and Keenan, D.J., *English Law*, 12th edn, Pitman Publishing Ltd, London (1998)
Walker, D.M., *The Scottish Legal System*, 7th edn, W. Green, Edinburgh (1997)
Walker, R.J., *The English Legal System*, 8th edn, Butterworths, London (1998)
Williams, Glanville, *Learning the Law*, 11th edn, Sweet and Maxwell, London (1982)

Law dictionaries

Curzon, *A Dictionary of Law* 5th edn, Pitman, London (1998)
Jowitt and Burke, *Dictionary of English Law*, 2nd edn plus supplement, Sweet and Maxwell, London (1995)
Mozley and Whiteley's Law Dictionary, 11th edn, Butterworths, London (1993)
Rutherford, L.A. and Bone, S., eds. *Osborn's Concise Law Dictionary*, 8th edn, Sweet and Maxwell, London (1993)
Concise Dictionary of Law, 4th edn, Oxford University Press, Oxford (1997)

Chapter 2
Principal health and safety Acts

S. Simpson

UK health and safety legislation consists of a number of main or principal Statutes or Acts which are supported by a great deal of subordinate legislation in the form of Regulations and Orders. This chapter deals with the more commonly applied main Acts that are concerned with protecting the health and safety of the working population and those who may be put at risk from the manner in which the work is carried out.

2.1 The Health and Safety at Work etc. Act 1974

2.1.1 Pre-1974 legislation

For more than a century health and safety legislation for persons at work in the UK had developed a piece at a time, each piece covering a particular class of person and not in a consistent manner each time. Separate legislation with variations in details and in the methods of enforcement would apply to a process or requirement when undertaken in a factory, as opposed to an office, a mine or a quarry. For example, an air receiver situated in a factory would be required to be examined for safety reasons by a competent person at least once every 26 months, but the same receiver moved to a shop would not require examination nor would the same receiver need to be inspected in the factory if, instead of air, another gas at the same or even higher working pressure was substituted.

In the main, the principal Act affecting the particular groups of persons, usually on the basis of the kind of premises in which they worked, was supplemented by regulations. The Act and its regulations would be enforced by a particular inspectorate (e.g. by factory inspectors for factories and notional factories such as construction sites, mines inspectors for mines and quarries and local authority inspectors for offices and shops). Any breach of the appropriate legislation could lead to a prosecution by an inspector which in turn could lead to a fine usually imposed on the company or other organisation rather than an individual.

The major responsibility for observing the requirements of the legislation was that of the employer with some responsibilities falling on the occupier, if he was not the employer, and on the employees. Only in mining legislation was there also a criminal liability placed on managers and other officials. On the whole, legislation tended to look to the protection of plant and equipment as a way of preventing injuries to workers. Visitors, contractors, neighbours and other third parties were mainly ignored in the drafting of these earlier Acts and regulations, as were many employees who did not work on premises (e.g. roadsweepers) or worked in premises not covered (e.g. schools, research establishments, hospitals, etc.).

By 1970 many organisations, especially the trade unions, were questioning whether the existing legislation was either sufficient or effective in providing proper protection for work people.

The effect that workers' organisations could have on workshop safety was limited and large sections of the working population were not covered.

A Private Member's Bill aimed at providing for the compulsory involvement of workers in accident prevention was withdrawn when in 1970 a committee was set up under the chairmanship of Lord Robens to look at safety and health at work. After studying the whole problem in depth the committee reported in 1972[1] making many recommendations of a wide ranging nature.

The essence of the 'Robens Report' recommendations was to:

1 Replace the mass of existing safety legislation with one Act applying generally to all persons at work.
2 Replace the mass of detail with a few simple and easily assimilated precepts of general application.
3 Change methods of enforcement so that prosecution is not always the first resort.
4 Ensure that occupational safety should also protect visitors and the public.
5 Place more emphasis on safe systems of work rather than technical standards.
6 Actively involve the workers in the procedures for accident prevention at their place of work.

In spite of changes of Governments, the main recommendations of the Robens Committee were accepted by Parliament and were incorporated in the Health and Safety at Work etc. Act 1974 (HSW).

2.1.2 The Health and Safety at Work etc. Act 1974

Drafted as an enabling Act, it permitted the Secretary of State or other Ministers to make regulations with a view to replacing the existing piecemeal legislation, typified by those Acts listed in schedule 1 of HSW, by regulations and codes of practice requiring improved standards of safety, health and welfare. It established a co-ordinating enforcement authority, the Health and Safety Commission (HSC), giving its inspectors greater powers than hitherto. It also extended legislative protection for health and safety to everyone who was employed, whether paid or not (except domestic servants), and imposed more general but wider duties on both employer and employee.

The Act makes provision for protecting others against risks to health and safety from the way in which work activities are carried out. It also seeks to control certain emissions into the atmosphere, as did the Control of Pollution Act 1974, and to control the storage and use of dangerous substances. In addition, the Act ensures the continuation of the Employment Medical Advisory Service.

Although gradually being superseded there is still a need to comply with the requirements of parts of the pre-1974 legislation which remain in effect but which apply only to those work activities covered previously.

2.1.3 General duties on employers and others

These duties are outlined in ss. 2–5 where the obligations are qualified by the phrases 'so far as is reasonably practicable' and 'best practicable means'. Interpretations of these phrases have been made[2] which indicate that 'reasonably practicable' implies a balance of the degree of risk against the inconvenience and cost of overcoming it, whereas 'best practicable means' ignores the cost element but recognises possible limitations of current technical knowledge.

In common law, employers have had, and still have, duties of care with regard to the health and safety of their employees, duties which are now incorporated into statute law as part of s. 2 of this Act.

The first part of s. 2 contains a general statement of the duties of employers to their employees while at work and is qualified in subsection (2) which instances particular obligations to:

1. Provide and maintain plant and systems of work that are safe and without risks to health. Plant covers any machinery, equipment or appliances including portable power tools and hand tools.
2. Ensure that the use, handling, storage and transport of articles and substances is safe and without risk.
3. Provide such information, instruction, training and supervision to ensure that employees can carry out their jobs safely.
4. Ensure that any workshop under his control is safe and healthy and that proper means of access and egress are maintained, particularly in respect of high standards of housekeeping, cleanliness, disposal of rubbish and the stacking of goods in the proper place.
5. Keep the workplace environment safe and healthy so that the atmosphere is such as not to give rise to poisoning, gassing or the encouragement of the development of diseases. Adequate welfare facilities should be provided.

In this section 'work' means any activities undertaken as part of employment and includes extra voluntary jobs for which payment is received or which are accepted as part of the particular job, i.e. part-time firemen, collecting wages etc.

Further duties are placed on the employer by:

s. 2(3) To prepare and keep up to date a written safety policy supported by information on 'the organisation and arrangements for carrying out the policy[7]. The safety policy has to be brought to the notice of employees. Where there are five or less employees this section does not apply.

s. 2(6) To consult with any safety representatives appointed by recognised trade unions to enlist their co-operation in establishing and maintaining high standards of safety.

s. 2(7) To establish a safety committee if requested by two or more safety representatives.

The general duties of employers and self-employed persons include, in s. 3, a requirement to conduct their undertakings in such a way that persons other than their employees are not exposed to risks to their health and safety. In certain cases information may have to be given as to what these risks are.

Landlords or owners are required by s. 4 to ensure that means of access or egress are safe for those using their premises and these are defined in s. 53 as any place and, in particular, any vehicle, vessel, aircraft or hovercraft, any installation on land, any offshore installation and any tent or movable structure. However, safety in workplaces, on vehicles etc. and on offshore installations are being overtaken by EU directives.

Those in charge of premises are required by s. 5 to use the best practicable means for preventing noxious or offensive fumes or dusts from being exhausted into the atmosphere, or that such exhausts are harmless. Offensive is not defined and may depend upon an individual's opinion.

Duties are placed by s. 6 on everyone in the supply chain, from the designer to the final installer, of articles of plant or equipment for use at work or any article of fairground equipment to:

1 ensure that the article will be safe and without risks to health at all times when it is being set, used, cleaned or maintained,
2 carry out any necessary testing and examination to ensure that it will be safe, and
3 provide adequate information about its safe setting, use, cleaning, maintenance, dismantling and disposal.

These duties are further extended in detail for machinery[3], electrical and electronic apparatus[4], gas appliances[5], lifts[6] and other equipment that is required to carry the CE mark before it can be put on the EU market.

There is obligation on designers or manufacturers to do any research necessary to prove safety in use. Erectors or installers have special responsibilities to make sure that when handed over the plant or equipment is safe to use. Obligations on designers are reinforced in regulations covering construction[7], offshore installations[8] etc.

Similar duties are placed on manufacturers and suppliers of substances for use at work to ensure that the substance is safe when properly used, handled, processed, stored or transported, to provide adequate information and do any necessary research, testing or examining. There are regulations detailing how substances and preparations should be classified, packaged and labelled with, in addition, the need for safety data sheets to be provided[9].

Where articles or substances are imported, the suppliers' obligations outlined above attach to the importer, whether a separate importing business or the user himself.

Often items are obtained through hire-purchase, leasing[10] or other financing arrangements with the ownership of the item being vested with the financing organisation. Where the financing organisation's only function is to provide the money to pay for the goods, the suppliers' obligations do not attach to them.

The employees' duties are laid down in s. 7 which state that, whilst at work, every employee must take care for the health and safety of himself and of other persons who may be affected by his acts or omissions. Also employees should co-operate with the employer to meet legal obligations. Section 8 requires that no one, whether employee or not, shall either intentionally or recklessly, interfere with or misuse anything, whether plant equipment or methods of work, provided by the employer to meet obligations under this or any other related Act.

The employer is not allowed by s. 9 to charge any employee for anything done or provided to meet statutory requirements.

2.1.4 Administration of the Act

The Act through s. 10 caused the establishment of two bodies to direct and enforce legislative matters concerned with health and safety. The Health and Safety Commission (HSC), appointed by the Secretary of State, consists of a chairman and six to nine members. Three of the members are appointed after consultation with the employers' organisations, three after consultation with employees' organisations and two after consulting local authorities.

It is the duty of the Commission (s. 11) to:

1 assist and encourage persons in furthering safety,
2 arrange for the carrying out of research and to encourage research and the provision of training and information by others,
3 provide an information and advisory service,
4 submit proposals for regulations, and
5 report to and act on directions given to it by the Secretary of State.

It also liaises with local authority and fire authority organisations to whom it has delegated[11,12] (s. 18) some of its duties.

Whereas the Commission has the function of formulating policies, the Health and Safety Executive (HSE) is responsible for their implementation. The Executive which is appointed by the Commission and consists of three persons, one of whom is the director, has a duty to exercise on behalf of the Commission such functions as the Commission directs. If so requested by a Minister, the Executive shall provide him with information of the activities of the Executive on any matter in which he is concerned and to provide him with advice.

The Commission may direct the Executive or authorise any other person to investigate or make a special report on any accident, occurrence, situation or other matter for a general purpose or with a view to making regulations.

The duties of the Commission and the Executive are contained in ss. 11–14 of HSW.

2.1.5 Regulations and Codes of Practice

The enabling powers of this Act are exercised through s. 15 whereby the appropriate Secretary of State or Minister may without referring the matter to Parliament require regulations to be drawn up by the Executive and submitted through the Commission to him. Such regulations may need to be submitted to Parliament for ratification. Although there is a general requirement for the Commission and Executive to keep interested parties 'informed of and adequately advised on, such matters' (s. 11(2)c) there is no obligation to consult. However, in drafting regulations that affect workplace safety, extensive consultation does occur.

The regulations may repeal or modify any of the existing regulations and matters related to ss. 2–9 of the Act. They can also approve or refer to specified documents, such as British Standard Specifications. A list of 22 subject matters that can be covered by regulations is given in schedule 3 of the Act.

The need to provide guidance on the regulations is recognised in s. 16 which gives the Commission power to prepare and approve Codes of Practice on matters contained not only in the regulations but also in ss. 2–7 of the Act.

To implement the EU framework and its daughter directives, a 'six-pack' of regulations was introduced in 1992 covering management[13], work equipment[14], display screens[15], manual handling[16], personal protective equipment[17] and health, safety and welfare[18]. The 'management' regulations extend HSW by requiring employers to:

- carry out risk assessments
- have arrangements for the planning and control of protective and preventive measures
- appoint competent persons to give health and safety assistance
- have procedures to cope with serious and imminent danger
- give information to employees
- co-operate and co-ordinate with other employers sharing the same premises
- take into account the employee's capabilities and training when entrusting tasks
- protect both young workers and pregnant workers

- give special consideration to workers who have recently given birth
- provide information to temporary workers.

Through the Fire Precautions (Workplace) Regulations 1997 these 1992 regulations encompass fire safety.

Before approving a code, the Executive acting for the Commission must consult with any interested body. The Commission have powers to approve codes prepared by bodies other than themselves, and some British and harmonised Standards have been approved.

An Approved Code is a quasi-legal document and although non-compliance with it does not constitute a breach, if the contravention of the Act or a regulation is alleged, the fact that the code was not followed will be accepted in court as evidence of failure to do all that was reasonably practicable. A defence would be to prove that something equally as good or better had been done (s. 17(2)). To supplement the Approved Codes of Practice, the Executive issue guidance notes which are purely advisory and have no standing in law.

2.1.6 Enforcement

2.1.6.1 General

The enforcement of the Act (s. 18), with some exceptions in respect of noxious and offensive emissions[19] (s. 5), is the responsibility of the HSE through its constituent inspectorates with certain premises delegated to local authorities[11] and for certain fire matters to the Fire Authority[12].

Actual enforcement is carried out by inspectors (s. 19) who should have suitable qualifications and be authorised by a written warrant outlining the powers they may exercise. An inspector must produce his warrant on request; without it he has no powers of enforcement.

2.1.6.2 Powers of inspectors

By virtue of his warrant an inspector has the powers outlined in s. 20 which relate only to the field of the inspectorate authorising him and include:

1. The right to enter premises and if resisted to enlist the support of a police officer.
2. To inspect the premises.
3. To require, following an incident, that plant is not disturbed.
4. Taking measurements and photographs although in the latter case it is usual to obtain permission first.
5. Taking samples of suspect substances.
6. Require tests to be carried out on suspect plant or substances.
7. Requiring the dismantling of plant.
8. Require those with possible knowledge relevant to his investigation to give it either verbally or in a written statement. The inspector has discretion to allow another to be present during questioning and the taking of a written statement.

9 The right to inspect and take copies of books or documents required to be kept by safety or other legislation if it is necessary for him to see them as part of his investigation but he has no right to examine documents for which legal privilege is claimed.
10 Requiring assistance within a person's limits of responsibilities.

Where an inspector takes samples of substances he must leave a similar identified sample with a responsible person or leave a conspicuous notice stating that he has taken a sample.

Information contained in an answer to an inspector cannot be used in criminal proceedings against the giver.

A customs officer may seize any imported article or imported substance and detain it for not more than two working days on behalf of an inspector (s. 25A).

Where an employer suffers damage to property or business, as a result of actions of an inspector, the inspector can be sued personally for recompense against which he may be indemnified by the enforcing authority.

After an inspector has completed his investigation he has a duty to inform representatives of the workpeople of actual matters he has found (s. 28(8)) and must give the employer similar information.

2.1.6.3 Notices

If an inspector is of the opinion that a breach has, or is likely to, occur he may serve an Improvement Notice (s. 21) on the employer or workman. The notice must state which statutory provision the inspector believes has been contravened and the reason for his belief. It should also state a time limit in which the matter should be put right.

However, if the activity involves immediate risk of serious personal injury, the inspector may serve a Prohibition Notice (s. 22) requiring immediate cessation of the activity. This notice must state what, in the inspector's opinion, is the cause of the risk and any possible contravention. If the risk is great but not immediate a deferred Prohibition Notice may be served stating a date after which the activity must cease unless the matter has been put right. Where corrective work cannot be completed in time, the inspector may extend the period of the notice. There is no procedure for certifying that a notice has been complied with.

Appeals against a notice may be made to an Industrial Tribunal[20]. On entering an appeal an Improvement Notice is suspended until the appeal is disposed of or withdrawn, whereas a Prohibition Notice continues in effect unless the Tribunal directs otherwise.

2.1.7 Offences

Offences listed in s.33 include:

1 failing to discharge a duty imposed by ss. 2–7,
2 contravening ss. 8 and 9, any regulation or notice,
3 making false entries in a register,

4 obstructing or pretending to be an inspector, and
5 making false statements etc.

If an inspector decides to institute legal proceedings, he must do so within six months of learning of the alleged contravention (s. 34(3)). Cases can be heard either summarily which attracts a fine not exceeding level 5 on the standard scale on conviction, or on indictment where the penalty can be imprisonment and/or an unlimited fine. Offences concerned with interfering with the powers or work of an inspector (s. 33(1)d,f,h and n) are to be dealt with summarily but for all the other offences listed in s. 33(1) plus in certain circumstances contravention of a requirement imposed by an inspector in the exercise of his powers (s. 33(1)e) the case can be tried either summarily or, if the offence is serious enough and the parties agree, on indictment, when the penalty on conviction can be an unlimited fine.

Responsibility for an offence usually attaches to the employer but may attach to an employee (ss. 7–8). However, where the contravention was caused with the consent or knowledge or be due to the neglect of a director, manager, company secretary or other officer (s. 37) then he too can be prosecuted.

In proceedings alleging a failure to use reasonably practicable or best practicable means the prosecution only has to state the suspicion and it is up to the accused to prove that what was done was as good as, if not better than, the duty required (s. 40).

Penalties have recently been increased by the Offshore Safety Act 1992 so that failing to discharge a duty under ss. 2–6 attracts a liability on summary conviction to a fine not exceeding £20 000 and on conviction on indictment to an unlimited fine. For specified offences, a person (such as a director, manager etc.) found guilty of the offence shall be liable on summary conviction to imprisonment, for a term not exceeding six months or a fine not exceeding £20 000 but for conviction on indictment, to imprisonment for a term not exceeding two years or a fine or both. Fines for other offences are set at level 5 (at present, through the Criminal Justices Act 1991, this is a sum not exceeding £5000).

2.1.8 Extensions

Part 1 of the Act has been extended to include:

1 the protection of the public from danger associated with the transmission and distribution of gas through pipelines,
2 securing the health, safety and welfare of persons on offshore installations engaged in pipeline works,
3 securing the safety of such installations and preventing accidents on or near them,
4 securing the proper construction and safe operation of pipelines and preventing damage to them,
5 securing the safe dismantling, removal and disposal of offshore installations or pipelines, and
6 the police.

2.1.9 Parts II to IV and Schedules

Part II of the Act allows for the continuation of the Employment Medical Advisory Service, defines the purpose and responsibilities of the service, allows for fees to be charged, for payments to be made and for the keeping of accounts.

Part III, except for s. 75, has been repealed by the Building Act 1984.

Part IV is a miscellaneous and general part amending the Radiological Protection Act 1970, Fire Precautions Act 1971, Companies Act 1967 and stating such matters as the extent and application of the HSW Act.

The following schedules of the Act cover:

1 Relevant existing enactments.
2 The constitution etc. of the Commission and Executive.
3 Subject matter of health and safety regulations.
4–7 Repealed.
8 Transitional provisions with respect to Fire Certificates.
9 Repealed.
10 List of repealed Acts.

2.1.10 Definitions

Sections 52 and 53 contain a number of definitions aimed at clarifying part I of the Act:

> ''Work' means an activity a person is engaged in whether as an employee or as a self-employed person. An employee is considered to be at work all the time he is following his employment whether paid or not and a self-employed person is at work throughout such time as he devotes to work as a self-employed person. Regulations can extend the meaning of 'work' and 'at work' to other situations such as YTS training[21].'

Other definitions include:

> ''Article for use at work' includes any plant designed for use at work and any article designed for use as a component in such plant.
> 'Code of practice' includes a standard, a specification or any other documentary form of practical guidance.
> 'Domestic premises' means premises occupied as a private dwelling (including gardens, yards, garages etc.).
> 'Employee' means an individual who works under a contract of employment.
> 'Personal injury' includes any disease or any impairment of a person's physical or mental condition.
> 'Plant' includes any machinery, equipment or appliance used at work.

'Premises' include any place, vehicle, vessel, aircraft, hovercraft, installation on land, offshore installation, installation resting on the sea bed or other land covered by water and any tent or movable structure within territorial waters. This definition has been extended by the Health and Safety at Work etc. Act 1974 (Application outside Great Britain) Order 1995 to include offshore installations, wells and pipelines, mines under the sea etc.

'Self-employed person' is an individual who works for gain or reward otherwise than under a contract of employment, whether or not he employs others.

'Substance' means any natural or artificial substance whether solid, liquid, gas or a vapour and includes micro-organisms.'

2.2 The Factories Act 1961

The Factories Act 1961 was in the main a consolidating Act, bringing together earlier Factories Acts. Most of the major provisions with regard to health, safety and welfare no longer continue in force.

However, those sections that do remain in effect refer to particular safety requirements but apply only to factories as defined in the Act and cover only hoists and lifts; chains, ropes and lifting tackle (soon to be replaced by the Lifting Operations and Lifting Equipment Regulations 1998 (LOLER)); prevention of explosions, and water sealed gas holders.

2.3 The Fire Precautions Act 1971

The Act furthers the provisions for the protection of persons from fire risks. If any premises are put to use and are designated, a certificate is required from the fire authority. Although classes of use cover the provisions of sleeping accommodation; use as an institution; use for the purposes of entertainment, recreation, instruction, teaching, training or research; use involving access to the premises by members of the public and use as a place of work, so far only the provision of sleeping accommodation and use as a place of work have been designated.

Houses occupied as single private dwellings are exempt, but the fire authority have powers to make it compulsory for some dwellings to be covered by a fire certificate.

Applications for fire certificates must be made on the prescribed form and the fire authority must be satisfied that the means of escape in case of fire, means of fire fighting and means of giving persons in the premises warning in case of fire are all adequate. Every fire certificate issued shall specify particular use or uses of the premises, its means of escape, details of the means of fire fighting, and of fire warning and, in the case of factories, particulars of any explosive or highly flammable materials which may be stored or used on the premises. The certificate may impose such restrictions as the fire authority considers appropriate and may cover the instruction or training of persons in what to do in case of fire or it may limit the number of persons who may be in the premises at any one time. In certain circumstances the fire authority may grant exemption from the requirements to have a fire certificate, otherwise a copy of the fire certificate is sent to the occupier and it must be kept on the premises. The owner of the building is also sent a copy of the certificate.

It is an offence not to have or to have applied for a fire certificate for any designated premises. Contravention of any requirement imposed in a fire certificate is also an offence. A person guilty of an offence (with some exceptions) shall be liable on summary conviction to a fine not exceeding level 5 on the standard scale and on conviction on indictment a fine or imprisonment or both.

So long as a certificate is in force, the fire authority may inspect the premises to ascertain whether there has been a change in conditions. Any proposed structural alterations or extensions to the premises, major changes in the layout of furniture or equipment or, in factories, to begin to use or store or increase the extent of explosive or flammable material shall, before the proposals are begun, be notified to the fire authority.

It is also necessary while the certificate is in force, or an exemption has been granted under s. 5A, for the occupier to give notice of any proposed material extension or alterations to the premises or its internal arrangement and, in the case of factories, to store or use or to materially increase the amount of explosive or highly flammable materials. Within two months of receiving notice, the fire authority must, if they regard the requirements of the relevant fire certificate as becoming inadequate, inform the occupier, or owner, and give such directions as they consider appropriate. If the directions are duly taken the fire authority will amend the certificate or issue a new one. Not giving suitable notice or contravening a direction are offences that on conviction could lead to a fine or imprisonment, or both. The rights of appeal are detailed in s. 9.

The coming into effect of the Fire Safety and Safety of Places of Sport Act 1987 amended but did not replace the FPA and gave the Fire Authority much wider powers. These include the power to charge a reasonable fee for the initial issue, or the amendment or the issue of a new fire certificate (s. 8B). Even though premises may be exempt from the requirements for a fire certificate there are duties to provide both means of escape and means of fighting fire (s. 9A). In order to assist occupiers to meet these duties the Secretary of State may issue Approved Codes of Practice and the fire authority may serve Improvement Notices if they think a code is not being met (ss. 9A–9F).

Should the fire authority be of the opinion that, in the event of fire, the use of premises involves or will involve so serious a risk to persons on the premises that continuing use ought to be prohibited or restricted, the authority may serve a Prohibition Notice on the occupier. There are rights of appeal against these notices (ss. 10–10B).

The Secretary of State has powers under the Act to make regulations about fire precautions in designated premises other than those in which manufacturing processes are carried on (s. 12). Requirements have been further extended by the Fire Precautions (Workplace) Regulations 1997 which apply particularly to premises for which a fire certificate is not required.

The Act deals with matters pertaining to building regulations (s. 13), the duties of consultation between local authorities (ss. 15 et seq.), fire authorities and other authorities such as the HSE, the enforcement of the Act (s. 18), the powers of inspectors to enter premises (s. 19), offences, penalties and legal proceedings (ss. 22–27) and the amendment of other Acts (ss. 29 et seq.).

Schedule 1 has the effect of making special provisions for factory, office, railway or shop premises, that do not form part of a mine, in relation to leasing, part ownership, the issue of licences under the Explosives Act 1875 and the Petroleum (Consolidation) Act 1928. It also has an effect on the proposed or actual storage or use of explosives or highly flammable material in factory premises.

2.4 The Mines and Quarries Acts 1954–71

The main Acts laying down the general safety duties of mines and quarries personnel (i.e. owners, managers, undermanagers, surveyors and officials) were the Mines and Quarries Act 1954, the Mines and Quarries (Tips) Act 1969 and the Mines Management Act 1971. The latter Act in particular and the mines sections of the 1954 Act were revoked by the Management and Administration of Safety and Health in Mines Regulations 1993.

Parts of the 1954 Act dealing with quarries have also been replaced by regulations.

2.5 The Environmental Protection Act 1990

To prevent the pollution from emissions to air, land or water from scheduled processes the concept of Integrated Pollution Control has been introduced. Authorisation to operate the relevant processes must be obtained from the enforcing authority which, for the more heavily polluting industries, is H. M. Inspectorate of Pollution. Control for pollution to air from the less heavily polluting processes is through the local authority.

Regulations also place a 'duty of care' on all those involved in the management of waste, be it collecting, disposing of or treating *Controlled Waste* which is subject to licensing.

In addition to extending the Clean Air Acts by including new measures to control nuisances, the Regulations introduce litter control; amend the Radioactive Substances Act 1960; regulate genetically modified organisms; regulate the import and export of waste; regulate the supply, storage and use of polluting substances and allow the setting up of contaminated land registers by the local authority. In 1991 the Water Act 1989 which controlled the pollution and supply of water was replaced by five separate Acts (see section 30.3.3).

Environment Agencies were set up by the Environment Act 1995 which also makes provision for contaminated lands, abandoned mines, control of pollution and the conservation of natural resources and the environment.

2.6 The road traffic Acts 1972–91

The road traffic Acts, including the Road Traffic Regulation Act 1984, together with the Motor Vehicle (Construction and Use) Regulations 1986, Road Vehicle Lighting Regulations 1981, Goods Vehicles (Plating and Testing) Regulations 1988, the Motor Vehicle (Tests) Regulations 1982 and numerous other regulations, form comprehensive safety legislation not only of the occupants of the vehicles but also for members of the general public who may be affected by the driving and parking of vehicles.

In the construction of vehicles, safety features include the provision of suitable braking systems; burst-proof door latches and hinges; material for fuel tanks; types of lamps and reflectors; the fitting of audible warnings, mirrors, safety glass windscreens, seat belts; acceptable tyres, the driver's view of the road and the lighting of vehicles. Noise and smoke emissions are also topics related to safety and covered by the legislation.

When loading a vehicle care must be taken to ensure that the load is evenly distributed to conform to the vehicle's individual axle weight and where necessary the driver must make suitable corrections on multi-delivery work to ensure that no axle becomes overloaded due to transfer of weight. Since it is an offence to have an insecure load, all loads must be securely fixed and roped and, if necessary, sheeted. Restrictions are placed on projecting loads, extra long or extra wide loads and abnormal indivisible loads. The carriage of dangerous goods, be they toxic, flammable, radioactive or corrosive, is covered by regulations made under other Acts (e.g. Petroleum (Consolidation) Act).

The road traffic Acts also deal with offences connected with the driving of motor vehicles and of traffic generally, accidents, road safety, licensing of drivers, driving instruction, restrictions on the use of motor vehicles, periodic testing of vehicles to ensure that they are roadworthy etc.

2.7 The Public Health Act 1936

This is another consolidating Act and in Part III statutory nuisances and offensive trades are dealt with.

Statutory nuisances are any premises in such a state as to be prejudicial to health or a nuisance, likewise the keeping of any animal, allowing any accumulation or deposit and causing any trade, business or process dust or effluvia to affect inhabitants of the neighbourhood. Not ventilating, not keeping clean and not keeping free from noxious effluvia or overcrowding any workplace are also statutory nuisances.

Where a statutory nuisance exists, the local authority can serve an abatement notice on the appropriate person, owner or occupier. If the abatement notice is disregarded, the court has powers to make a nuisance order.

Consent of the local authority is required before specified trades or business can be carried on. These offensive trades include blood boiling and drying, bone boiling, fat extracting and melting, fell mongering, glue making, soap boiling, tripe boiling and dealing in rags and bones.

The Act gives local authorities power to make bye-laws with regard to offensive trades and of fish-frying.

An allied piece of legislation is the Food Safety (General Food Hygiene) Regulations 1995 made under the Food Safety Act 1990. These regulations apply in England and Wales only but similar food hygiene regulations also exist for Scotland and for Northern Ireland.

The principal requirements of the Regulations relate to:

(a) the cleanliness of premises and the equipment used for the purpose of a food business;
(b) the hygienic handling of food;
(c) the cleanliness of persons engaged in the handling of food;
(d) the construction of premises used for the purposes of a food business and their repair and maintenance;
(e) the provision of water supply and washing facilities;
(f) the proper disposal of waste;

2.8 Petroleum (Consolidation) Act 1928

Little remains of this Act which is now restricted to licensing for the keeping of petroleum spirit, the making of byelaws for filling stations and canals and to testing petroleum. Extant regulations cover compressed gas cylinders, the keeping of petroleum spirit and extending the provisions of the Act to other substances such as carbide of calcium and liquid methane.

2.9 Activity Centres (Young Persons Safety) Act 1995

With the growth of centres providing facilities where children and young persons can engage in adventure activities, and as a result of tragedies due to poor management of such centres, the need to control these centres became apparent. The Act allows for the making of regulations such as those prescribing the type of person who should hold a licence to provide and run the centre, the duties of the licensing authorities and enforcement of the Act[22,23].

2.10 Crown premises

Although s. 48 of the HSW makes provision for binding the Crown to the provisions of part of the Act it excepts ss. 21 to 25 which deal with prohibition and improvement notices. The Crown immunity exists because in the exercise of justice in the name of the monarch it is not constitutionally possible for one part of the Crown service to pursue another part of the service into the courts. Nevertheless, the HSE does apply a version of prohibition and improvement notices, called Crown Notices, when deficiencies are found on Crown property. Through the National Health Service (Amendment) Act 1986, Crown immunity for both food legislation and health and safety legislation have been removed from the Health Authority.

2.11 Subordinate legislation

This chapter has dealt briefly with the major legislation concerned with health and safety which is becoming progressively more proscriptive, i.e. it states the aims to be achieved but not how to achieve them. Supporting this major legislation is a vast and growing body of subordinate legislation, mainly in the form of regulations, which derive their authority from the main Acts. While not being completely proscriptive, these regulations, which tend to occur in particular areas of health and safety concern, point to possible routes to compliance while recognising that there may be other equally effective means for achieving the same goal.

The range of these regulations is continually developing and the list that follows the references gives some of the more generally applicable examples. Many of them are referred to in the texts of the following chapters where their relevant content is discussed.

References

1. Report of the Committee on Safety and Health at Work. 1970–71 (Robens Report), Cmnd 5034, HMSO, London (1972)
2. Fife, I. and Machin, E.A. *Redgrave Fife and Machin Health and Safety*, Butterworth, London (1990)
3. Health and Safety Executive, *The Supply of Machinery (Safety) Regulations 1992*, SI 1992 No. 2063, HMSO, London
4. Health and Safety Executive, *The Electromagnetic Compatibility Regulations 1992*, SI 1992 No. 2372, HMSO, London
5. Health and Safety Executive, *The Gas Appliances (Safety) Regulations 1995*, SI 1995 No. 1629, HMSO, London
6. Department of Trade and Industry, *The Lifts Regulations 1997*, SI 1997 No. 831, HMSO, London
7. Health and Safety Executive, *The Construction (Design and Management) Regulations 1994*, SI 1994 No. 3140, HMSO, London
8. Health and Safety Executive, *The Offshore Installations and Wells (Design and Construction etc.) Regulations 1996*, SI 1996 No. 913, HMSO, London
9. Health and Safety Executive, *The Chemicals (Hazard Information and Packaging for Supply) Regulations 1994*, SI 1994 No. 3247, HMSO, London
10. Health and Safety Executive, *The Health and Safety (Leasing Arrangements) Regulations 1992*, SI 1992 No. 1524, HMSO, London
11. Health and Safety Executive, *The Health and Safety (Enforcing Authorities) Regulations 1989*, SI 1989 No. 1903, HMSO, London
12. Health and Safety Executive, *The Fire Precautions (Factories, Offices, Shops and Railway Premises) Order 1989*, SI 1989 No. 76, HMSO, London
13. Health and Safety Executive, *The Management of Health and Safety at Work Regulations 1992*, SI 1992 No. 2051, HMSO, London
14. Health and Safety Executive, *The Provision and Use of Work Equipment Regulations 1992*, SI 1992 No. 2932, HMSO, London
15. Health and Safety Executive, *The Health and Safety (Display Screen Equipment) Regulations 1992*, SI 1992 No. 2792, HMSO, London
16. Health and Safety Executive, *The Manual Handling Operations Regulations 1992*, SI 1992 No. 2793, HMSO, London

17. Health and Safety Executive, *The Personal Protective Equipment at Work Regulations 1992*, SI 1992 No. 2966, HMSO, London
18. Health and Safety Executive, *The Workplace (Health, Safety and Welfare) Regulations 1992*, SI 1992 No. 3004, HMSO, London
19. Health and Safety Executive, *The Health and Safety (Emission into the Atmosphere) Regulations 1983*, SI 1983 No. 943, HMSO, London
20. Health and Safety Executive, *The Industrial Tribunals (Improvement and Prohibition Notices Appeals) Regulations 1974*, SI 1974 No. 1925, HMSO, London
21. Health and Safety Executive, *The Health and Safety (Training for Employment) Regulations 1990*, SI 1990 No. 1380, HMSO, London
22. Health and Safety Executive, *The Adventure Activities (Licensing) Regulations 1996*, SI 1996 No. 772, HMSO, London
23. Health and Safety Executive, *The Adventure Activities (Enforcing Authorities and Licensing Amendment) Regulations 1996*, SI 1996 No. 1647, HMSO, London

List of some generally applicable regulations

The Highly Flammable Liquids and Liquefied Petroleum Gases Regulations 1972
The Safety Representatives and Safety Committee Regulations 1977
The Control of Lead at Work Regulations 1980
The Health and Safety (First Aid) Regulations 1981
The Control of Industrial Major Accident Hazards Regulations 1984
The Classification and Labelling of Explosives Regulations 1983
The Asbestos (Licensing) Regulations 1983
The Ionising Radiations Regulations 1985
The Control of Asbestos at Work Regulations 1987
The Electricity at Work Regulations 1989
The Pressure Systems and Transportable Gas Container Regulations 1989
The Noise at Work Regulations 1989
The Health and Safety (Training for Employment) Regulations 1990
The Control of Explosives Regulations 1991
The Lifting Plant and Equipment (Records of Test and Inspection etc.) Regulations 1992
The Management of Health and Safety at Work Regulations 1992 as variously amended
The Provision and Use of Work Equipment Regulations 1992
The Manual Handling Operations Regulations 1992
The Workplace (Health, Safety and Welfare) Regulations 1992
The Personal Protective Equipment Regulations 1992
The Health and Safety (Display Screen Equipment) Regulations 1992
The last six regulations are referred to as the 'six-pack'.
The Supply of Machinery (Safety) Regulations 1992
The Electromagnetic Compatibility Regulations 1992
The Ionising Radiation (Outside Workers) Regulations 1993
The Control of Substances Hazardous to Health Regulations 1994
The Construction (Design and Management) Regulations 1994
The Gas Safety (Installations and Use) Regulations 1994
The Gas Appliances (Safety) Regulations 1995
The Chemical (Hazard Information and Packaging for Supply) Regulations 1994 as amended in 1996 and 1997
The Reporting of Injuries, Diseases and Dangerous Occurrences Regulations 1995
The Food Safety (General Food Hygiene) Regulations 1995
The Health and Safety (Signs and Signals) Regulations 1996
The Adventure Activities Licensing Regulations 1996
The Adventure Activities (Enforcing Authority and Licensing Amendment) Regulations 1996
The Gas Safety (Management) Regulations 1996
The Health and Safety (Consultation with Employees) Regulations 1996
The Work in Compressed Air Regulations 1996
The Carriage of Dangerous Goods (Classification, Packaging and Labelling) and Use of Transportable Pressure Receptacles Regulations 1996

The Construction (Health, Safety and Welfare) Regulations 1996
The Health and Safety (Young Persons) Regulations 1997
The Lifts Regulations 1997
The Confined Spaces Regulations 1997
The Fire Precautions (Workplace) Regulations 1997
The Diving at Work Regulations 1997
The Confined Spaces Regulations 1997

Further reading

Selwyn, N., *Law of Health and Safety at Work*, Croners, London (1993)
Mahaffy and Dodson on Road Traffic, Butterworth, London, (loose-leaf)
Garner, *Environmental Law*, Butterworth, London, (loose-leaf)
Encyclopedia of Environmental Health, Sweet and Maxwell, London (loose-leaf)

Chapter 3
Influences on health and safety
J. R. Ridley

3.1 Introduction

Laws don't just happen.

They are a reaction to a perceived need, whether real or imaginary, and are aimed at protecting the individual and/or the community as a whole. But within the perceived need there are many trends and factors that influence attitudes towards the sort of action that needs to be taken. With time, situations and attitudes – both personal and political – change, making what was right yesterday not necessarily right today. These changes and developments are no more visible than in health and safety, an area on which a great deal of political, community and media attention has been focused and which can become very emotive issues. This chapter looks at some of the influences that affect the current laws, attitudes, interpretations and practices in health and safety.

3.2 The Robens Report[1]

There can be no doubt that the single largest influence on the organisation of and our approach to health and safety lies with the work of the Robens Committee and its subsequent report. Their remit gave them the opportunity to stand back and take an objective look at where health and safety was going – not only in the UK but in other major manufacturing countries. Their considerations were carried out in a political atmosphere that was changing from employer dominated to one in which the employee had a much greater say. Added to this, employees were no longer the down-trodden, exploited victims of a culture that blossomed in an environment of extremes of poverty and riches, a culture that had precipitated the passing of the first ever law to protect working people.

All the recommendations contained in the Robens Report were accepted by the government of the day and by the opposition parties. They have since been incorporated into health and safety legislation, notably the Health and Safety at Work etc. Act 1974 (HSW). Major changes brought about as a result include:

- all employees except domestic servants were brought within the scope of the Act and given a measure of protection while at work
- laws have been made more flexible to enforce and to comply with
- making laws less prescriptive and more proscriptive
- simplifying the process for making subordinate legislation
- increasing the powers of inspectors to reduce some of the bureaucratic and time-consuming enforcement procedures
- giving Minsters of State powers to approve codes of practice whose contents showed a means of achieving conformity with the laws
- accepting the concept of putting responsibility where authority lay, i.e. with the employer and of the need for a commitment at senior levels in an organisation
- recognising that employees were not the only ones who had a right to protection from the effects of the way work was carried out
- confirming the role that employees have to play in health and safety
- advocating a greater degree of self-regulation through the use of own rules and codes of practice.

It is over 25 years since the Roben's committee carried out its review of health and safety in the UK. In that time a great many changes have occurred in the health and safety field, changes in the attitudes of employers, employees and society as a whole, in the origins of laws and in the emphasis of legislative requirements. This raises a question of whether the time is approaching when a further review is needed to build on some of the radical changes initiated by the Roben's Report. Some relevant issues are considered in the following sections of this chapter.

3.3 Delegation of law-making powers

The original laws of the country, or dooms as they were then called, were made by the king. As the national administration became more complex, the law-making powers were delegated to, then finally taken over by, Parliament. Up until fairly recently, the laws were debated in depth by Parliament who incorporated a great deal of detail into their content so they stated not only what was to be achieved but how. Inevitably with the increasingly complex nature of the laws required, this proved a ponderous and time-consuming activity. So in the early part of the twentieth century the practice developed of Parliament delegating that part of its law-making authority concerned with the minutiae of compliance to a Minister of the Crown.

With the enormous strides in the development of technology that have been made over the past few decades the practice of delegating powers for making subordinate laws has become the norm in respect of health and safety. These powers are given only to the Secretary of State for the particular government department concerned (s. 15 HSW) who may require one of his Ministers to oversee the preparation of the necessary documents. As far as health and safety is concerned the drafting of proposals for subordinate legislation is normally passed to the HSC who, in turn, pass to the HSE the actual drafting work. Section 15 of HSW states '... the Secretary of State and that Minister acting jointly shall have power to make regulations ...'. The interpretation of the word 'make' is important. It does not mean that they can put laws onto the Statute Book but only that they can prepare (make) the documents that state the requirements. To give the proposals the power of law needs the authority and approval of Parliament, which it gives through the procedure described below.

In the process of drafting its proposals, the HSE is required to consult with '... any government department or other body that appears ... to be appropriate ...' (s. 50 HSW). The procedure followed by the HSE is similar to that followed by Parliament with white and green papers (see section 1.15.3) where there may be consultation with the CBI, TUC and an industry with a particular interest in the subject. Thereafter public consultation takes place through the publication of a Consultative Document (CD) which outlines and explains the new proposals.

Opinions submitted are taken into account in the preparation of the final proposal which is then sent to the HSC for its approval. From there it goes to the Minister concerned who takes it to Parliament for their approval. Before that approval can be given, the proposal must lie in the House for thirty days during which time any MP can peruse it, raise objections and call for a debate on it.

At the end of the thirty days a vote is taken on whether to approve the proposal or reject it. There are two procedures for voting on a regulation 'made' by a Minister. The first for proposals of minor legal importance covering such administrative matters as the timing of the implementation of parts of an Act, or the approval of certain types of safety equipment where a *negative* vote is taken, i.e. if there are no objections to the proposal its approval is assumed and it becomes part of the law. The second procedure is followed for proposals that are of greater importance, such as those that may qualify or change part of an existing Act where a *positive* vote is required, i.e. there must be a majority vote in favour of the proposal for it to become part of the statute law. With subordinate legislation there is no need for royal assent since that is inherent in the authority contained in the substantive Act.

3.4 Legislative framework for health and safety

The foundations of health and safety legislation are the relevant Acts passed by Parliament on the subject – most noticeably HSW. But this particular Act covers such a wide range of subjects and activities that it needs supportive subordinate legislation to put flesh on its legislative skeleton. HSW incorporates clauses that empower the appropriate Secretary of State and/or the Minister to 'make' regulations and orders.

These regulations have the objective of detailing legislative requirements in the specific area at which they are aimed although there has been a trend – reflecting the requirements of EU directives – of being more proscriptive in the wording of the clauses or regulations than previously and of relying on annexes or schedules to identify areas that require particular attention. Conformity with the legislative requirements is then measured against compliance with particular official codes of practice or standards, whether BS or transposed harmonised (BS EN) standards.

To support the requirements, and to strengthen the arm of the enforcing authorities, HSW allows the HSC, subject to the approval of the Secretary of State, to make and approve Codes of Practice whether of their own drafting or of others. The fact of giving official approval to Codes of Practice (ACoPs) gives them a status of recognition in law such that they will be accepted by a court as documentary evidence of a means of conforming with the legislative requirements without the need to 'prove' the document should the defendant challenge it.

In the 'making' of Codes of Practice for approval there are statutory requirements that HSC/E should consult with *'any government department or other body that appears . . . to be appropriate . . .'* (s. 16(2) HSW). The use of ACoPs is becoming increasingly common as new regulations – reflecting EU directives – become more proscriptive. ACoPs lend greater flexibility to the making and amending of statutorily recognised standards – this is an important aspect that allows rapid changes to be made to keep pace with new hazards introduced by rapidly advancing technologies.

While ACoPs fill an important role they do not completely fill the gap which shows acceptable means for achieving conformity with legislative requirements – this is left to various advisory publications issued by HSC/E, – from the L series of booklets that explain the legislative requirements through the HS(G) guidance booklets to the series of technical Guidance Notes covering chemical, environmental hygiene, medical, plant and machinery aspects of health and safety.

In addition, a number of industries, often working with the HSC's Industry Advisory Committees (IACs), draw up codes of practice in respect of hazards particular to their operations. Where these are recognised or published by HSE they provide a statement of ways in which conformity with legislative requirements can be achieved.

3.5 Self-regulation

Self-regulation is an ideal to be aimed for where organisations set their own effective standards, make rules to ensure those standards are met and enforce adherence to those rules. The political mood of the 1970s and 1980s, looking for means to reduce the legislative burden on industry, turned to self-regulation as an answer. However, they failed to recognise that in the smaller companies, which comprise by far the greater number of employers, the top priority was economic survival in the face of increasing economic strictures, with health and safety as an also-ran. Furthermore, the attitudes and cultures to make such an innovation a success did not exist – nor was the political will sufficiently persistent to carry it through.

The move towards self-regulation reflects the political drift of the 1980s/90s to reducing the burden of imposed controls on employers and giving them greater freedom to act within their particular employment circumstances and environment. While many major manufacturing companies have practised this sort of control for years, it was from the consumer market that the motivation came. Those areas of industry and commerce that provide goods and services to major companies who supplied the consumer market were required to control the quality of goods to the standards demanded by major buyers. They had set their own standards and demanded it of their suppliers, in respect not only of quality but of health, safety and hygiene in the manufacture of those goods. This practice is likely to grow as the trend spreads to the manufacturing sector and more retail organisations demand quality assurance on goods they sell as 'own brand'.

Areas where self-regulation is included in legislation are seen in the requirements of both the Pressure Systems Regulations[2] and Lifting Operations and Lifting Equipment Regulations 1998[3] (LOLER) whereby employers and users decide what examinations need to be carried out and at what intervals according to the circumstances of use of the equipment.

For self-regulation to be a success requires:

- a commitment at the most senior level in an organisation
- a culture pervading all levels of the organisation that will accept only the highest standards (much as exists in successful quality assurance systems)
- a commitment by and the active co-operation of all levels
- the will to make the necessary resources available.

The advocation of self-regulation must not be a reaction to the excessive zeal or over-dogmatic enforcement of statutory requirements but must stem from a genuine desire to achieve high standards in ways that are pertinent to the particular organisation. The desire to implement schemes for self-regulation must be generated within the organisation. A move to encourage businesses in this direction can be seen in the HSC's campaign 'Good health is good business'.

The success of self-regulation depends to a large extent on the honesty and integrity of those following its regime. There is something of a parallel in the self-certification of machinery allowed under the Machinery Directive[4] but this requires the support of documentary evidence which can be examined.

3.6 Goal-setting legislation

Effectively this was first introduced by HSW which laid down broad objectives to be achieved rather than the very prescriptive nature of earlier legislation with its roots in a manufacturing-led economy. Now that manufacturing is no longer the major employer, and is facing severe economic and competitive strictures, the imposition of unrealistic controls is more likely to accelerate its demise than encourage its survival. To survive and expand requires a regime where controls are related to operating circumstances rather than to broad brush catch-all requirements. Legislation needs to recognise this and be more flexible in its ability to accommodate the enormous range of employment conditions and circumstances met today while still ensuring that acceptable standards are maintained. This requires a more flexible proscriptive approach which allows individual industries and employers to determine how best, within their trading and employment situation, they can achieve the desired end results.

Where industries and companies do agree and set standards, the role of enforcement – whether internally or by HSE inspectors – should be to ensure those standards are met and maintained.

The principle of goal setting permeates the making of EU health and safety directives and stems from a Resolution adopted by the European Council in June 1985 on a 'new approach to legislative harmonisation' whereby goal-setting objectives are outlined in broad terms in the main Articles of the directives and rely on supportive annexes to specify the particular areas requiring detail consideration. For directives concerned with work equipment, the annexes include a list of Essential Health and Safety Requirements (ESRs). In turn the matters covered by the annexes rely on harmonised standards to give evidence of conformity with the ESRs and hence the directive.

Goal-setting legislation cannot stand by itself but must rely heavily on associated requirements, in the form of Approved Codes of Practice, official guidance material, national and harmonised standards etc. to provide a base against which the degree of conformity can be assessed.

3.7 European Union

When the UK became a member state of the European Economic Community in 1972 it agreed to be bound by the legislative procedures of the Community (now the European Union (EU)). At that stage each of the Member States had the power of veto over any proposed EEC legislation with the result that very little progress was made towards unified laws. That changed in 1986 with the adoption of the Single European Act[5] (SEA) which brought in qualified majority voting (QMV), a system whereby each Member State is allocated a number of votes weighted according its population. *Table 3.1* shows the allocation of votes to the various Member States with a summed total of 87 votes. A majority vote (a minimum of 62) is needed for a matter to be adopted. This means that the UK lost its power of veto over directives and has become subject to the political inclinations and standards of its European neighbours with their cumulative majority vote.

The EU is a complex organisation which contains a number of salient institutions with various executive and advisory roles.

The *Council* is made up of the representatives of the governments of the 15 Member States. Normally these are Ministers of State with the Foreign Secretary being the UK's main representative. However, Ministers from other Departments meet to deal with specific matters such as finance, agriculture, transport etc. Twice a year the Heads of State meet for a *European Council* (sometimes referred to as the Summit). The Council is

Table 3.1 Allocation of 'weighted' votes to member states

Member State	Number of votes
France	10
Germany	10
Italy	10
United Kingdom	10
Spain	8
Belgium	5
Greece	5
Netherlands	5
Portugal	5
Austria	4
Sweden	4
Denmark	3
Finland	3
Ireland	3
Luxembourg	2
Total	87

Minimum number of votes needed for adoption of a proposal = 62

the supreme law-making body of the EU and is the only body that can adopt EU legislation. A great deal of background work is carried out between Council meetings by the *Committee of Permanent Representatives* whose members are the ambassadors to the EU of each member state.

The *Commission* is the executive body of the EU and consists of 17 members appointed by agreement between member governments. Commissioners must act independently of their national government and of Council. The Commission is answerable to the European Parliament and its detailed work is carried out by 23 Directorate Generals (DGs) each dealing with a specific aspect of EU business. Those most relevant to health and safety are:

DG III – Internal market and industrial affairs – covering the construction of work equipment for free movement in the EU.

DG V – Employment, social affairs and education – dealing with safety in the workplace and the safe use of work equipment.

DG XI – Environment, consumer protection and nuclear safety.

The Council and Commission are assisted by the *Economic and Social Committee* (EcoSoC) who must be consulted before decisions are taken on a large range of subjects. It is non-governmental and consists of 186 members representing the social partners plus particular sector interests. Additionally, preliminary proposals with a health and safety content are referred to the *Advisory Committee on Safety, Hygiene and Health Protection at Work* for their opinion.

The *Court of Justice*, which sits in Luxembourg, consists of 13 judges appointed by agreement of Member States. Its role is to hear cases concerning EU legislative matters. It does not interfere with national judiciary other than to advise where a national matter impinges on or clashes with EU law.

The *Court of Auditors* oversees the use of EU funds and ensures the proper use of such monies.

The *European Parliament* (EP) consists of 626 members, each elected by universal suffrage, who attach to EU-level political groups – there are no national sections. Detail consideration of specific matters is delegated to one of 18 standing committees whose rapporteurs report their findings to EP. There is an extensive consultation procedure between the Council, the Commission and EP on proposed legislation but EP seem anxious to increase their involvement in EU law making.

A *European Agency for Health and Safety at Work* has been established to gather and disseminate information on health and safety matters to all member states. The Agency is located in Bilbao.

Each of these various institutions has a particular and separate function within the EU but work closely together to make an effective organisation.

3.7.2 EU legislation

Within the EU there are four basic types of legislation:

- *Regulations* which have direct applicability in Member States and take precedence over national laws. They arise mainly in respect of the coal and steel industries. They are rarely used for health and safety matters but have been used on transboundary environmental matters.
- *Directives* are the more usual EU legislation for health and safety issues. They do not have direct applicability but put an onus on Member States to incorporate their contents into national laws within a time scale specified in the directive.
- *Decisions* and *Recommendations* relate to specific matters of local concern and apply directly to the Members States at whom they are directed.

When a directive is adopted by the Council of Ministers, each Member State becomes committed to incorporate into its national laws the contents of the directive. This means in a number of areas – such as health and safety, employment, environment, finance etc. – the UK Parliament's power to decide the subject matter of the laws it passes is being eroded. While there are still domestic matters on which Parliament can decide to legislate, there is a growing body of laws whose subject matter and content have been dictated to Parliament. As the EU gets stronger and the eurocrats more bureaucratic the question arises of how much longer can the UK retain some measure of right of sovereignty in the making of its laws.

In the health and safey field, the EU approach to law making has polarised into two discrete areas. For new equipment, health and safety has been used as the criteria for setting equipment construction standards to allow free movement throughout the EU. This concerns free trade and is dealt with by Directorate General III. However, the use of work equipment – whether existing or new – is considered to be part of employment and is dealt with by Directorate General V. In both cases, early drafts of proposals are referred to the *Advisory Committee on Safety, Hygiene and Health Protection at Work* for its opinion. This committee is tripartite in composition with representatives from each Member State and can be a useful forum for influencing the final proposals in favour of national and sectorial positions.

One of the moves to create a free European market which allows free movement throughout the EU for all new machinery requires that the machinery conforms to the Machinery Directive[4]. Similar moves aimed at ensuring a greater degree of uniformity of working conditions in Member States requires comformity with the Workplace Directive[6] and Work Equipment Directive[7]. These directives were adopted in 1989 and had to be implemented in member states by 1 January 1993, the date when the European free market became established. In the UK the directives concerned with work equipment have been transposed into the Supply of Machinery (Safety) Regulations 1992 (SMSR) and the Provision and Use of Work Equipment Regulations 1992 (PUWER) which are considered in detail in later chapters.

3.8 European standards

Harmonised or 'EN' standards are those that have been approved by the European standard making organisations, Committé European de Normalisation (CEN) for standards of mechanical equipment and Committé European de Normalisation Electrotechnique (CENELEC) for standards of electrical equipment. They are given the prefix letters EN and in the UK can be recognised by the designation BS EN, e.g. BS EN 836, *Garden equipment – Powered lawn mowers – Safety.*

These standards are prepared by working parties whose members are representatives of the participating Member States – including members of EU, EFTA plus some from Eastern European and Middle East States. Representation at CEN and CENELEC is through the various national standards making bodies – in the UK, BSI.

There are parallel international standards making bodies, International Standards Organisation (ISO) for mechanical standards and International Electrical Commission (IEC) for electrical standards. These European and international standards making bodies are now working closely together to prevent duplication of effort and to expedite the preparation of standards.

3.8.1 Harmonised standards making procedures

In the UK, standards are developed by the British Standards Institution which remains the national standards making body but, with membership of the EU, procedures for making standards changed. In 1983 the

Commission reached agreement with the two European standards making bodies that conformity with harmonised standards would be accepted as evidence of conformity with the directives. These two bodies, CEN and CENELEC, are sponsored by the Commission but act independently.

As soon as a subject is selected for a harmonised standard all Member States must cease individual work in the area. The work of drafting a harmonised standard is carried out by working groups of representatives of the Member State national standards making bodies. Final drafts are circulated to all members of CEN/CENELEC to vote for approval or rejection. A majority vote is required before the standard is adopted. Once adopted, the EN standard then applies in all EEA Member States and takes precedence over any national standard that covers the same subject.

Subjects for an EN standard are determined from a perceived need or by a proposal from a Member State that is considered to be of benefit to the EU as a whole. Where a Member State has a standard which they consider could beneficially apply across the EU they can submit it as the basis for an EN standard. BS 5304 'Safety of machinery'[8] was put forward and accepted on this basis. In the event, rather than keeping it as a comprehensive standard on the guarding of machinery, it has been split into component subjects each of which has become a separate harmonised standard. With the adoption of these EN standards, BS 5304 has had to be withdrawn as a British Standard and declared obsolete. However, because it is so comprehensive, well understood and widely used, it is likely to continue to be recognised in the UK as giving a standard suitable for conformity with PUWER

In their approach to developing harmonised standards, CEN and CENELEC have categorised standards into four types:

Type A – generic safety standards covering basic concepts, principles for design and general aspects that can be applied to all machinery.

Type B – dealing with one safety aspect or one type of safety related device that can be used across a wide range of machinery:
- type B1 cover particular safety aspects such as safety distances, surface temperatures, noise etc.
- type B2 cover safety related devices such as two hand controls, interlocking devices, pressure sensitive devices, guards etc.

Type C – giving detailed safety requirements for a particular machine or group of machines – machine specific standards.

The category status of harmonised standards is quoted in the introduction to each standard.

3.9 Our social partners

A phrase that has become part of the jargon of the EU and which effectively refers to the relationship between employer and employee. Its meaning has been enlarged from a definition of a simple relationship to encompass a very complex inter-relationship and in so doing its influence has grown enormously, particularly in the field of employment and health and safety. Employers' and employees' representatives have had a running dialogue for many years, going back to the days of the big unions and national wage negotiations. More recently big has become no longer beautiful and negotiations on conditions of work have become much more localised. But there is a continuing dialogue between employer and employees.

However, the attitudes, practices and relationships in our European partners with their influences through the European Commission and Parliament are resulting in changes to working standards in the UK. The EU works by consensus or through qualified majority rule recognising the standards of the less advanced Member States in the development of EU-wide standards. In effect, EU legislation sets the highest acceptable standard then allows derogation for those Member States that are less technicologically advanced. If the EU legislative requirements are less than those currently accepted in the UK, the question arises of how the UK can maintain its safety standards without its industries suffering economically from the resultant higher manufacturing costs.

3.10 Social expectations

Over the past two or three decades, the social attitude to work has undergone a considerable change. It has moved from one of being glad to have a job (and an income) and accepting the rough with the smooth where the rough could be very rough and result in horrendous injuries for which little or no compensation was paid. Also wages were such that there was little enough to barely survive on. Hours were long and the work demanding which left little time or energy for anything outside the work. There was little social life other than the men congregating in the pub or the women gossiping over the back fence. Rarely was there social entertainment for a man and his wife and, indeed, their family except on feast days and at the travelling fairs. Life centred on work and that was the sole object in life. However, with the considerable rise in incomes and living standards that have occurred since the end of the Second World War radical changes in social behaviour have occurred. Workpeople have the leisure time and are able to afford an extensive social life often making greater inroads on their time and energy than their paid employment. With these changes in circumstances has come a change in attitude towards the dangers experienced at work. Workpeople now expect to be able to enjoy their leisure time and demand that any dangers from the work process be removed and that employers provide conditions of work that do not put them – the workpeople – at risk.

3.11 Public expectations

Public expectations in health and safety are largely determined by the way in which a particular risk is viewed. The response to a hazard or risk, whether at individual, group or national level is significantly influenced by how it is perceived. To illustrate the point, each year many thousands of people are killed or injured on the roads and this fact rarely gets reported in the press. However, a multi-vehicle crash or a coach crash which causes several deaths immediately reaches the front page of the national press and television. In the area of health, whilst many people die from heart disease and cancers often related to smoking or excessive alcohol consumption, the media ignores them and focuses on the small number who die from the newly highlighted disease, Creutzfeldt–Jacob Disease (CJD). The occurrence of multi-death events or strange new diseases attract the media whose reporting treatment causes greater concern in the minds of individuals and the general public than the number of victims would seem to warrant.

Work by Slovic, Fischhoff and Lichenstein[9] shows that people are frequently unaware of the true risks from a specific hazard. For example, their study showed that whilst the actual number of deaths in the USA from botulism was less than 10 per year, it was estimated at between 100 and 5000 by the representative group of interviewees. Similarly the group estimated that deaths from stomach cancers at 5000 as against an actual figure of 100 000 per annum. The authors noted that biases in newspaper coverage closely matched the biases in people's perception. The question is, did the researchers report a natural bias or had the newspapers created one?

Further work by the same researchers indicated that different societal groups have different perceptions. In a survey, college students, members of a professional and business organisation and members of a women's political organisation all ranked nuclear power among the riskiest of 30 activities, whilst professional risk assessors placed it a lowly twentieth. Nuclear power is a very highly regulated activity and has become so because the media focus and public perception combined to provoke strong regulatory responses from the law makers.

Overall, the less that is known about a recognised risk, the more it is feared and the greater the anxiety about it. Conversely, the more we are familiar with a risk and feel in control of it, the less concerned we are about it. The concern arising from the perception of a risk can be at an individual, local or national level. When it surfaces at national level it frequently results in regulatory controls.

3.12 Political influences

These really operate at two levels: the grass roots level where the work people and electors put pressure on their Parliamentary representatives to make laws that provide protection against perceived dangers, conversely, politicians are anxious to appease the electorate and keep their votes so take up the rifles loaded by their constituents and fire the bullets. In some cases these bullets hit the target as Private Member's, Bills and partisan enactments are passed.

At other times an event or incident can occur whose result is so horrifying that even Parliament is jolted into action. Unfortunately there seems to be chronic shortsightedness among MPs with the result that in these circumstances any legislations passed tends to be aimed only at the circumstances and industry involved in the incident and to miss entirely the wider opportunity that the situation offers.

Finally there is the great annual party pilgrimage where the ruling party of the day decides in its infinite wisdom, and with an eye on political expediency, to introduce legislation that will increase its standing as a political party among marginal groups of voters. With luck, and it sometimes happens, the subject chosen will be in need of legislative control and the working community as a whole will benefit, but it is a random chance because the motivation behind it is fed by a desire to make political gain. Rationale does not seem to enter the issue when political standing is at stake.

3.13 Roles in health and safety

Early legislation put the responsibility for remaining safe at work very firmly in the hands of those who *operated* the machinery, so that if they had an accident it was their fault and they got no compensation. However, the pendulum swung very much the other way in the mid parts of this twentieth century with the onus being put firmly on the shoulders of the employer in their position of *controlling* what was done and how. But the pendulum is swinging back with the growing realisation of the positive role that employees have to play in health and safety. The catalyst for this oscillation seems to be *knowledge* which is being imparted through the extensive training programmes within industry and the higher level of education generally in the community at large. It remains to be seen how long it will be before and to what extent the judiciary will acknowledge and recognise that those with knowledge have an obligation to employ it in meeting statutory obligations.

Some of this has already been recognised in law through the obligations placed on manufacturers and suppliers to pass to their customers information about hazards and safe operating techniques for the equipment and substances that they supply. This obligation can be met relatively easily, and seen to be met, by preparing and issuing written instructions and data sheets. What is not going to be so easy to demonstrate is the extent of take-up of knowledge by an employee during training. Will NVQs and SNVQs have a role to play in providing documentary evidence of the knowledge of the workpeople? And to what extent will that documentary evidence of *qualification* be recognised by the courts as imposing a legal obligation on individual workpeople to work safely and following working rules?

3.14 Safety culture

Anyone, who in the course of their work, has to visit a number of different workplaces very quickly acquires the facility for getting the 'feel' of the atmosphere in the workplace. That feel can be confirmed in the state of the workshops, offices and particularly the toilets which give an indication of the attitude of the workpeople. In any normal community there are inevitably one or two disgruntled individuals with genuine or imagined complaints against society. But when a whole community in a workplace exhibits this attitude the cause needs to be sought in a factor common to that workplace and that single common factor is often the manager, director or owner. It is a strange quirk of organisation that the attitude held in the board room, although there is no direct communication, inevitably manifests itself in the attitudes and behaviour of the shop floor or in the general office. This attitude or culture, which permeates the organisation, emanates from the highest level.

The implications of this in health and safety are enormous. It's a bit like the adage 'look after the pennies and the pounds will look after themselves'. If the attitude at the top of an organisation is concerned with achieving high standards of health and safety that attitude will permeate the organisation and be measurable in the working areas with high levels of safety performance. Closely associated with this will also be high levels of job satisfaction with the spin-off of high quality of product and high output.

Thus the safety culture, which in itself cannot be quantified or evaluated as a function in absolute terms, can have great benefits to the organisation, its viability and to the people working in it. That safety culture can only be generated by the most senior people in that organisation – the board of directors – and this is probably the greatest contribution they can make towards the cause of high standards of health and safety.

3.15 Quality culture

There has been, in recent years, a recognition that to remain viable and to retain customers, the quality of product or service must be of an acceptably high standard. Much of the pressure for this has come from the need to remain competitive in a free market, but a great deal also stems from major purchasers, particularly in the retail and consumer durable markets, seeking greater market share through the quality of the goods and services they sell. They do this by ensuring within their own organisations high levels of service for the customer and by demanding from their suppliers a similar high quality in the goods they purchase. This latter point enables them to put enormous pressure on the manufacturer or producer to achieve the required standards. In many cases the purchaser will instruct the supplier in the techniques for achieving those quality standards. This move has been recognised nationally and internationally by the promulgation of standards[10,11]. These standards require systems to be in place that assure the quality of the items from the very start of manufacturing, and is based on the inculcation of a suitable attitude throughout all levels of the manufacturing organisation to achieve the goal.

So what is so different with health and safety? It requires the same attitude of mind and can bring similar benefits to the organisation. If the incentive is there for generating an attitude for satisfying a customer on quality, why should a similar attitude for protecting the most important and expensive asset in an organisation be so frequently lacking?

3.16 No fault liability

The concept of no fault liability is encompassed to a very low degree in the social security legislation[12] that ensures that anyone injured at work, subject to certain restrictions, can obtain a measure of financial payment while incapacitated. However, it is doubtful if that level of benefit is sufficient to sustain a standard of living equivalent to that enjoyed on full wages. In the event of ill-health or injury resulting from a work activity, compensation can still be sought from the employer in litigation with the surety that if damages are awarded payment will be guarateed under the compulsory insurance required by the Employer's Liability (Compulsory Insurance) Act 1969 and Regulations[13]. However, this can be a time-consuming and expensive process which most injured persons cannot afford unless supported by a union or other association.

In 1978 the Pearson Commission[14] made recommendations concerning the limited application of a system of no fault liability operating through a national insurance scheme. Because of sector interest pressure on the government of the day, nothing was done. However, schemes do exist in New Zealand and Canada which have been running for many years with varying degrees of success.

The question to be asked in respect of no fault liability, where individual employers contribute to a national scheme according to their performance, is whether the remoteness of the penalty (premium payment) from the events (the accidents and compensation payments) demotivates employers from taking actions necessary to remove the cause of the accident. This is a complex area since it inevitably involves humanitarian attitudes as well as the purely economic.

3.17 Conclusion

Activities, standards and legislation in the health and safety field are subject to an immense range of influences, at individual, company, community and national level. Standards resulting from community and national influences tend to be imposed and can be resented.

Of all these influences, probably the most effective is in the area of company activities where a close relationship between employer and employee can generate immediate and long-lasting working methods of a high safety standard. This results from the attitude and culture within the organisation over which the organisation has some control. However, this is not a one-way arrangement but requires an ongoing close co-operation between employer and employee to ensure its continuing success.

References

1. HM Government, *Report of the Roben's Committee*, Cmnd 5034, The Stationery Office Ltd, London (1972)
2. *The Pressure Systems and Transportable Gas Container Regulations 1989*, The Stationery Office Ltd, London (1989)
3. *The Lifting Operations and Lifting Equipment Regulations 1998*, The Stationery Office Ltd, London (1998)
4. European Union, Council Directive No. 89/392/EEC, *on the approximation of the laws of Member States relating to machinery* (Machinery directive), EU, Luxembourg (1989)
5. HM Government, *Single European Act 1986*, Cmnd 9758, The Stationery Office Ltd, London (1986)
6. European Union, Council Directive No. 89/654/EEC, *concerning the minimum safety and health requirements for the workplace*, EU, Luxembourg (1989)
7. European Union, Council Directive No. 89/655/EEC, *concerning the minimum safety and health requirements for the use of work equipment by workers at work*, EU, Luxembourg (1989)
8. British Standards Institution, BS 5304, *Safety of Machinery*, BSI, London (1988). This standard has been overtaken by harmonised EN standards and is now obsolete
9. Slovic, P., Fischoff, B. and Lichenstein, S., *Perceived Risk: Psychological Factors and Social Implications*, Proceedings of the Royal Society, London, A 376, 17–34 (1981)
10. British Standards Institution, BS 9001, *Quality systems – Specification for design/development, production, installation and servicing*, BSI, London (1994)
11. British Standards Institution, BS ISO 14001 *Environmental management systems. Specification with guidance for use*, BSI, London (1996)
12. *Social Security (Industrial Injuries and Diseases) Miscellaneous Provisions Regulations 1986*, The Stationery Office Ltd, London (1986)
13. *Employer's Liability (Compulsory Insurance) Act 1969*, The Stationery Office Ltd, London (1969)
14. HM Government, *Report of the Royal Commission on Civil Liability and Compensation for Personal Injury* (The Pearson Report), The Stationery Office Ltd, London (1978)

Chapter 4

Law of contract

R. W. Hodgin

4.1 Contracts

4.1.1 Formation of contract

A contract is an agreement between two or more parties and to be legally enforceable it requires certain basic ingredients. It must be certain in its wording and consist of an offer made by one party which must be accepted unconditionally by the other (*Scammell* v. *Ouston*[1]; *Carlill* v. *Carbolic Smoke Ball Co.*[2]). This does not prevent negotiations taking place and alterations being made by both parties during the early stages of the discussions, but in its final stage there must be complete and clear agreement as to the terms of the contract (*Bigg* v. *Boyd Gibbins Ltd*[3]). The great majority of contracts need not be in writing and those made daily by the general public, buying a newspaper or food, clearly show this.

There must, however, be 'consideration' that flows from one party to the other. Consideration is the legal ingredient that changes an informal agreement into a legally binding contract. It is the exchange of goods for payment, of work for wages, of a journey for the price of a ticket that amounts to consideration (*Dunlop Pneumatic Tyre Co. Ltd* v. *Selfridge & Co. Ltd*[4]). It is possible for future promises, mutually exchanged, to amount to consideration. Thus X agrees to buy a car from Y on January 1st and payment will not be made until that date. The contract dates from the day that these promises are exchanged.

Two other essentials for a valid contract are that the parties must intend to enter into a legally binding agreement and both parties must have the legal capacity to make such a contract. Lack of intention is rarely a problem in reality, but if one party alleges that he never intended to enter into a contract, only the courts can say, having analysed the behaviour of the parties, whether or not a contract was created (*Balfour* v. *Balfour*[5]). Capacity to contract means that the parties must be sane, sober and over the age of 18, although there are obvious exceptions to the age requirement when it comes to contracts of employment; such employment contracts are now largely regulated by statute, e.g. Employment of Children Act 1973.

Although the limits of a contract should be reflected in what the parties to it have expressly agreed upon, it is possible for the courts, or for Parliament, to imply terms into the agreement. The most notable example of statutory implied terms is the Sale of Goods Act 1979 which among other things implies into contracts of sale a condition that goods shall be of merchantable quality. The court's role in implying terms is a more difficult area of the law because the general approach adopted by the courts is one of non-intervention.

It is clear, however, that in certain circumstances they will adopt a more positive role particularly where the implied term is necessary to give business sense to the agreement (*Matthews* v. *Kuwait Bechtal Corporation*[6], *The Moorcock*[7]).

4.1.2 Faults in a contract

Despite the outward appearance of agreement there may be fundamental faults that will affect the validity of the contract. The parties are not always clear and precise in the language they use and it may be that the contract will be void, that is unenforceable, on the grounds of Mistake. This can arise in a number of ways. It may be that the subject matter of the contract is no longer in existence at the time the contract is made, e.g. where X appears to sell a machine to Y but earlier the machine had been destroyed in a fire at the factory.

It is possible that the parties have been negotiating at cross-purposes, e.g. where X intends to sell one particular machine but Y has in mind another machine at X's factory. In this situation the basic requirement of agreement is missing and no contract comes into existence.

A third possibility is where one party is mistaken as to the identity of the other contracting party and can prove that identity was crucial to his entering into the contract. It should be stressed, however, that the courts will not easily allow the mistaken party to avoid the contract for this would be an easy way for people to escape from contracts that look as though they are about to take a disastrous financial turn.

It may be that one party has been led into the contract by a Misrepresentation made by the other party, e.g. untrue statements about the capabilities of the machine. The remedies available in this situation vary depending on whether the misrepresentation was made innocently, negligently or fraudulently. Under the Misrepresentation Act 1967 where the statement was made innocently the party misled may ask for the contract to be set aside, but the court has the power to refuse this and instead grant damages. If the statement was made negligently then damages can be awarded and the contract may be set aside, but the court may decide that damages alone are sufficient and rule that the contract should continue. Lastly, the most serious misrepresentation is where it is fraudulent and here both damages and setting aside the contract will be ordered.

Even though the formation of the contract meets all the requirements it may still be declared to be an illegal contract and unenforceable. This is a complex topic but one example can be taken from the contract of employment. There are often restraint clauses in such contracts whereby an employer seeks to prevent an employee who leaves the firm from working for a rival company or setting himself up in competition. Such restraints are basically unenforceable because they are not in the public interest and are contrary to the employee's freedom to work. However, it is possible to enforce such a restraint if the employer can show that the wording of the clause was reasonable in scope and that he has some interests, such as trade secrets or customer lists, that need protecting (*Fitch* v. *Dewes*[8]). What the employer cannot do, however, is to seek to prevent competition that the ex-employee may threaten.

Similar restraints are often found in the sale of businesses. Part of the price of a business sale reflects the goodwill that the owner has built up over the years. The buyer obviously would not want the seller to start up in competition with him by opening a similar venture in a location which would pose a financial threat to the newly acquired business. The court approaches the problem in the same way as in employer–employee restraints. Basically the restraint will be struck out as being against the public interest unless the buyer can so word the restraint that the court regards it as reasonable in the circumstances. In both types of restraint the type of work or business, the length of time of the restraint and the geographical area of the restraint will all be taken into account by the court in deciding whether it is reasonable or not (*Nordenfelt* v. *Maxim Nordenfelt*[9]).

4.1.3 Remedies

In the vast majority of cases contracts are satisfactorily concluded with both sides completing their respective obligations. But when one party fails to do so and is in breach of contract then the question of remedies arises. The normal remedy is damages, or monetary compensation.

The aim of damages is to put the innocent party in the position he would have been in if the breach had not occurred (*Parsons* v. *B.N.M. Laboratories Ltd*[10]). It is not for him to profit from the wrongdoer's behaviour and there is in fact a duty on the innocent party to mitigate the loss wherever possible (*Darbishire* v. *Warren*[11]). The claim will also be limited to what the wrongdoer can reasonably have been expected to foresee would be the outcome of his breach. For instance where one party sends a piece of machinery to the other party for repair and the repairer is in breach of contract by not returning it by an agreed date, claims covering loss of production will be allowed if the repairer should have foreseen the likely losses caused by his delay (*Hadley* v. *Baxendale*[12]). The safest thing is to inform the repairer at the time the contract is made of the exact function the machinery plays in the manufacturing process so that he is aware of its importance (*Victoria Laundry (Windsor) Ltd* v. *Newman Industries Ltd*[13]).

Another possible remedy is *quantum meruit*. This arises where the innocent party has completed part of his contract but is prevented from continuing by the other party. His claim is then based on the amount of work he has completed up to that date (*Planche* v. *Colburn*[14]). However, this claim cannot be maintained by a party whose failure to complete is of his own doing. A builder cannot build half a garage and refuse to complete and yet claim for the work done. If the work is completed but badly, then the contract price must be paid less a deduction to compensate for the faulty work.

It is possible for the court to grant Specific Performance as a remedy whereby one party is ordered to complete his part of the contract. The remedy is discretionary and little used outside of land sales. It will not be granted where the contract is one of personal services, e.g. in a contract of employment. Although an industrial tribunal may order reinstatement of an employee following an unfair dismissal, such a remedy cannot be enforced against an unwilling employer and his refusal will merely be reflected in the compensation awarded to the former employee.

The above general discussion is obviously of the briefest nature. What follows is a closer look at specific contracts; contracts which depend for their content and form on legislation.

4.2 Contracts of employment

It is important to distinguish between contracts of service and contracts for services. The former describes the relationship between employer and employee while the latter is concerned with employing independent contractors to carry out certain specific tasks. Unfortunately it is not always easy to distinguish between the two and yet it is essential in order to determine the legal liabilities and responsibilities of the parties. This is particularly important in situations involving main contractors and sub-contractors. The wording of the contract can place responsibility on any party but care should be taken to set this out clearly in the various contracts. If this is done then the parties involved can cover their responsibilities by obtaining insurance.

In a contract of service it is said that a man is employed as part of the business; whereas under a contract for services his work, although done for the business, is not integrated into it but is only accessory to it (*Stevenson, Jordan and Harrison* v. *Macdonald and Evans*[15]; *The Ready-mixed Concrete (South East) Ltd.* v. *The Minister of Pensions and National Insurance*[16]). The distinction has serious repercussions on tortious liability for the general rule is that the employer is liable for the torts committed by his employees acting in the course of their employment, but he is not liable for the tortious behaviour of independent contractors. It must be stressed, however, that there are a number of exceptions to this basic rule. Even where an exception applies and the employer is liable to third parties, it may be that the contract will give the employer rights of reimbursement from the contractor.

4.3 Employment legislation

A contract of employment can be in any form, but the more informal it is the more difficult it may be to define its true scope.

Parliament enacted the first Contract of Employment Act in 1963, requiring that certain basic ingredients be expressed in writing. Governments since have been active in the area of employment law and much of the present law is to be found in the Employment Rights Act 1996 and the Trade Union Reform and Employment Rights Act 1993. The information to be communicated to the employee must be in writing and include the names of employer and employee, the date of commencement of employment, hours of work, pay, holiday entitlements, incapacity for work, sick pay provisions, length of notice which the employee must give and is entitled to receive, pension provisions and the employee's job title. Any changes in the terms of employment must be notified to the employee within one month of the change although they need not be retained by the employee.

Information regarding disciplinary rules and grievance procedure must also accompany the written particulars. These requirements are not, however, conclusive evidence of the terms of the contract of employment, but an employee can ask for the contract to be altered to correspond with the terms if he feels there are discrepancies. It is common also for the particulars to refer to other documentation, for instance, collective agreements, and by so doing to incorporate them into the Contract of Employment (*Systems Floors (UK) Ltd.* v. *Daniel*[17]). In all these the written agreement is persuasive but not necessarily conclusive evidence of the relationship between the parties.

The principal Act regulating employees' rights and employers' duties is now the Employment Rights Act 1996. This Act is a consolidation of earlier enactments and deals with such matters as pay, Sunday working, maternity rights, termination of employment and remedies for unfair dismissal.

Just a few of these points can be touched on in this chapter. An employee has a right to be paid in a situation where the employer has no work for him that day. To be eligible the employee must have been continuously employed for not less than one month. There will be no such entitlement if the lack of work is caused by industrial action nor will there be any entitlement if the employer has offered alternative and suitable work which has been refused. The amount of any guaranteed payment shall not exceed £14.50.

The Act guarantees the employee's right of membership of trade unions and also payment for time off work while participating in certain union or public duties. Where an employee is given notice of dismissal by reason of redundancy he is entitled to take reasonable time off work to look for new employment or to make arrangements for training and he is entitled to be paid for such absences. Provisions are made for time off work to attend antenatal care and for payment to cover such periods. Rights to maternity leave and pay are set out in detail in Part VIII of the Act.

An employee who is suspended from work by his employer on medical grounds is entitled to wages for up to 26 weeks. Such suspension must arise from a requirement imposed by law or under a recommendation in a Code of Practice issued under HSW in relation to the Control of Lead at Work Regulations 1980, the Ionising Radiations Regulations 1985 and the Control of Substances Hazardous to Health Regulations 1994. There is no entitlement, however, if the employee is incapable of work due to disease or bodily or mental impairment.

An employer must give a minimum period of one week's notice to terminate employment of less than two years; one week's notice for each year of employment up to 12 years of continuous employment and not less than 12 weeks' notice for a continuous work period in excess of 12 years. An employee must give to his employer a minimum of one week's notice.

Part X deals with the question of unfair dismissal while Chapter II deals with employees' rights when unfairly dismissed. Of particular interest is s. 100 entitled 'Health and safety cases'. This states that where an employee has been designated by the employer to carry out activities in connection with preventing or reducing risks to health and safety at work and he is dismissed for carrying out those duties, then such dismissal will be regarded as unfair. A similar attitude is adopted in relation to a situation where the employee is a member of a safety committee and is dismissed for carrying out his functions. Where an employee is dismissed for refusing to return to a place of work on the grounds that he reasonably believes it to be a danger then any dismissal will also be regarded as unfair.

Several other pieces of legislation also play an important role in determining the relationship between employer and employee. The Equal Pay Act 1970 (see also the Equal Pay (Amendment) Regulations 1983 introducing the concept of equal value) introduces into all contracts an equality clause stating that where people are employed on broadly similar work with people of the opposite sex then any discrepancy in terms or conditions between them must be equalised upwards unless such differences can be explained on grounds other than a difference of sex. The Sex Discrimination Act 1975 as amended by the Employment Act 1989 and the Race Relations Act 1976 render it unlawful for a person to treat another less favourably on sexual or racial grounds. Any employment variations must be shown to be justified for non-sexual or non-racial reasons.

Apart from the above legislation certain other terms are implied into the employment contract, having been built up by court decisions over the years. These can be supplanted only by express provisions of an Act so stating.

The most important of the implied conditions are:

- that both parties exercise reasonable care in carrying out their duties under the contract. This means that the employer should provide a safe system of work including machinery, a safe place of work and skilled fellow employees;
- the employee must take care to act reasonably and not injure others (*Lister* v. *Romford Ice and Cold Storage Co. Ltd*[18]);
- the employee owes his employer a duty of fidelity which prevents him from working for a rival firm or divulging secret information (*Hivac Ltd* v. *Park Royal Scientific Instruments Ltd*[19]);

- the employee should co-operate fully with his employer to achieve the goal of the employment contract;
- an employer will not conduct his business in a manner likely to destroy or seriously damage the relationship of confidence and trust between him and his employee (*Malik v. BCCI*[20]).

4.4 Law of sale

The most common example of a contract is one for the sale and purchase of goods. While the basic common law rules and rules of equity still apply to such sales, many of the rules are to be found in the Sale of Goods Act 1979, as amended, which consolidates legislation that began in 1893. The Act covers the whole range of contract topics such as formation, terms, performance, transfer of ownership, rights of unpaid sellers and remedies.

Worthy of particular mention here are the implied terms found in ss. 12–15. Implied terms, as we have seen in the previous section, are terms which the courts will read into the contract where parties have failed to mention them. Because of the hesitancy of the courts in implying terms into contracts, the 1979 Act specifically sets out the important terms that must be read into the contract of sale. These terms provide the buyer with a certain basic protection against buying faulty or unsuitable goods.

Section 12(i) states that there is an implied term that the seller has good title to the goods and is therefore capable of passing true ownership to the buyer (*Rowland v. Divall*[21]).

Section 13 covers a situation where goods are sold by description. There is an implied term that the goods will correspond to the description given (*Beale v. Taylor*[22]).

Section 14 contains two important implied terms where the seller is selling in the course of a business, in contrast to a private sale. The first is that the goods must be of a satisfactory quality unless the seller has drawn the buyer's attention to the defect or the buyer has examined the goods and should have detected the faults before the contract was made (*Wilson v. Rickett, Cockerell & Co. Ltd*[23]). The second implied term is that where the buyer has expressly or by implication made known to the seller the particular purpose for which the goods are bought then such goods must be reasonably fit for such purpose (*Henry Kendall & Sons v. William Lillico & Sons Ltd*[24]).

Section 15 is concerned with sales by sample and implies a term that the bulk will correspond to the sample in quality; that the buyer will have a reasonable opportunity of comparing bulk with sample and that the goods will be free from any defect rendering them unsatisfactory which would not have been apparent on examination of the sample (*Ashington Piggeries Ltd v. Christopher Hill Ltd*[25]).

Certain transactions, i.e. exchange and barter, whereby goods change ownership may not fall within the definition of 'sale of goods'. Also, where a repair is carried out, the transfer of any new part itself is not regarded as a sale of goods.

To give protection to the new owner the Supply of Goods and Services Act 1982 was introduced whereby similar implied terms to those listed above under the Sale of Goods Act 1979 are incorporated into the contract. As the title of the Act suggests it also covers the service element of the contract. Thus, where the supplier is acting in the course of business there is an implied term that he will exercise reasonable care and skill. Where no time is stipulated for completion of the service then there is an implied term that it must be carried out within a reasonable time and that is determined by the facts of the case. Likewise, where no price is fixed by the contract then there is an implied term that a reasonable price will be paid.

This is an opportune moment to mention one of the most important pieces of consumer legislation in recent years, the Unfair Contract Terms Act 1977. The original Sale of Goods Act 1893, following the general principles of the common law contracts, permitted the parties to exclude themselves from legal liability for wrongful performance of their contractual duties by suitably worded clauses or notices. The 1977 Act, together with the 1982 Act, now forbids such clauses any legal recognition if the aim is to avoid liability for injury or death caused by negligence, for instance where an electrical appliance is faulty.

Where the clause is aimed at avoiding economic loss caused by one party to the other, for instance by selling unmerchantable goods, the Act recognises two situations. If the seller sells in the course of business and the buyer is not buying in the course of business and the goods are of a type normally supplied for private use or consumption, then the Act prohibits the exclusion of the implied terms of the 1979 Act. Where the contract is not a consumer sale, then an exclusion clause will be valid but only if it can meet the test of reasonableness as laid down in the 1977 Act, the burden of proof resting on the party who wishes to utilise the exclusion clause (*George Mitchell (Chesterhall) Ltd.* v. *Finney Lock Seeds Ltd*[26]). Section 12 of the 1979 Act cannot be excluded in either category of sale. The 1982 Act adopts similar procedures. Somewhat surprisingly the 1979 Act does not apply to insurance contracts, an area where consumer protection would seem to be a necessary requirement.

The omission from the legislation was allowed in exchange for the insurance industry, through the Association of British Insurers, drawing up their own Statements of Insurance Practice. The purpose of such Statements is to give the private insured greater rights against his insurance company. The Statements do not, however, carry the force of law and represent only voluntary agreements on the part of those insurers who are members of A.B.I.

Also of great importance in terms of consumer protection is the Unfair Terms in Consumer Contracts Regulations 1994 which came out of an EU directive. As the title indicates, it is limited to consumer contracts only but it does apply to insurance contracts. It does not, however, apply to employment contracts. The determining factor is whether there is an 'unfair term'. This is defined as any term which, contrary to the requirements of good faith, causes a significant imbalance in the parties' rights and obligations under the contract to the detriment of the consumer. In assessing this requirement of good faith regard will be had to the strength of the bargaining positions of the parties and the extent to which the seller or supplier had dealt fairly and equitably with the consumer.

4.5 Specialised legislation affecting occupational safety advisers

The responsibility of occupiers of land to those who enter the premises is to be found in the Occupier's Liability Acts of 1957 and 1984. The 1957 Act covers both tortious and contractual liability. The basic obligation is that the occupier owes a common duty of care to see that the premises are reasonably safe for the purpose for which the visitor has been permitted to enter (s. 2). It is possible, of course, for any contract that may exist between the occupier and the visitor to state a higher duty of care than that of s. 2. But under the Act it was possible for the occupier to exclude this basic duty by a suitably worded exclusion clause in a contract or by a notice (*Ashdown* v. *Samuel Williams & Sons*[27]). But the Unfair Contract Terms Act 1977 now prohibits any attempt to avoid liability for personal injury or death caused by negligence where the premises are used for business purposes and the injured person was a lawful entrant. Even under the 1957 Act where a party entered premises by virtue of a contract to which he was not a party, for example contractor's workmen, any exclusion of liability in that contract did not affect his rights against the occupier. He was owed the common duty of care unless the contract stated that higher obligations were owed, in which case he could then benefit from the higher duty (s. 3).

The Occupier's Liability Act 1984 makes two important changes to the law. The original Act made no reference to the duty of an occupier to a trespasser. The courts were left to evolve their own rules to cater for this category of person.

Section 1 of the 1984 Act states that an occupier owes a duty to take such care as is reasonable in the circumstances for the safety of a trespasser if he is aware or should have been aware of the existence of the danger and if he knows or should know that a trespasser may come within the vicinity of the danger.

Section 2 of the Act makes a change to the Unfair Contract Terms Act so that where an occupier allows someone to enter his business premises for purposes that are recreational or educational and not connected with the business itself then the occupier may rely on the use of an appropriately worded exclusion clause or notice.

Special reference should be made to two sections of the Health and Safety at Work etc. Act 1974 in a chapter on contract.

Section 4 is concerned with the various duties owed by those who have control of premises to those who are not their employees. Subsection (3) states that where such a person enters non-domestic premises by virtue of a contract or tenancy which creates an obligation for maintenance or repair or responsibility for the safety of or absence from risks to health arising from plant or substances in any such premises, then the person deemed to be in control owes a duty to see that reasonable measures are taken to ensure that such premises, plant or substances are safe and without risks to the health of the person entering. This section therefore would provide that safety standards be extended to someone who enters a cinema (contract) or enters a factory to inspect machinery (licensee) in addition to the other aim of the Act which is concerned with the safety of employees. It should be noted that liability is on the person who has control over premises or who can be described as an occupier and that case law shows that more than one person can be in that position (*Wheat v. Lacon & Co. Ltd*[28]).

The Consumer Protection Act 1987, schedule 3 amends and widens the scope of s.6 of the Health and Safety at Work etc. Act 1974 which is concerned with the general standard of care and safety owed by manufacturers, designers, importers and suppliers of articles for use at work. Such persons must ensure, so far as is reasonably practicable, that the article is so designed and constructed that it will be safe and without risk to health at all times when being set, used, cleaned or maintained by a person at work. To meet these requirements there is a duty to carry out or arrange for such testing or examination as may be necessary in the circumstances. Also adequate information must be given to the person supplied about the use for which the article was designed. Revisions of earlier information must also be given. Similar obligations exist where it is a 'substance' rather than an 'article' that is being supplied. The major alteration here, however, is that the duties are not restricted to 'for use at work' but cover also when it is being 'used, handled, processed, stored or transported by a person at work'. It must be stressed that liability under the HSW Act 1974 is penal in character and civil remedies are only permissible where the Secretary of State introduces specific regulations.

However, there may be a contractual relationship between the supplier and the recipient of articles or substances referred to above. A remedy for breach of that contract may therefore be available. Subsection (8) of the HSW Act (as amended by the CP Act 1987) allows the originator of a defective article to escape liability if he has obtained a written undertaking from the person supplied that the person will take specified steps sufficient to ensure, so far as is reasonably practicable, that the article is safe and without risk to health when used, set, cleaned or maintained by a person at work. Subsection (9) also has the effect of pinpointing responsibility for contravening the standards imposed by s.6 on the effective supplier of such goods when another person has in fact been the contracting party with the customer by virtue of a hire-purchase agreement, conditional sale or credit sale agreement. Usually the financing of these arrangements is carried out by means of finance houses. Contractually the goods are sold to the finance house who then in turn enters into his own contract with the customer. This subsection rightly seeks to ensure that the basic obligations of s.6 remain with the originator of the faulty design, product etc., rather than allowing it to pass to the finance house. Similarly, the Health and Safety (Leasing Arrangements) Regulations 1980 extend to those who, purely as financiers, lease articles for use at work, the immunity from the duties of care imposed by s.6 and leave the obligation of this section on the shoulders of the effective supplier rather than the ostensible supplier.

Further reading

For an introduction to the law of contract see:
Davies on Contract (Upex), 7th edn, 1995 Sweet & Maxwell, London (1995)
For more detailed coverage of employment law see:
Pitt, *Employment Law,* 3rd edn, Sweet & Maxwell, London (1997)
Selwyn, N., *Law of Employment,* 9th edn, Butterworth, London (1996)
Cases and Material on Employment Law, 2nd edn, Blackstone (1995)

Reference cases

1. Scammell *v.* Ouston (1941) 1 All ER 14
2. Carlill *v.* Carbolic Smoke Ball Co. (1893) 1 QB 256
3. Bigg *v.* Boyd Gibbins Ltd (1971) 2 All ER 183
4. Dunlop Pneumatic Tyre Co. Ltd *v.* Selfridge & Co. Ltd (1915) AC 847
5. Balfour *v.* Balfour (1919) 2 KB 571
6. Matthews *v.* Kuwait Bechtal Corporation (1959) 2 QB 57
7. The Moorcock (1886–1890) All ER Rep 530
8. Fitch *v.* Dewes (1921) 2 AC 158
9. Nordenfelt *v.* Maxim Nordenfelt Guns and Ammunition Co. (1894) AC 535
10. Parsons *v.* B. N. M. Laboratories Ltd (1963) 2 All ER 658
11. Darbishire *v.* Warren (1963) 3 All ER 310
12. Hadley *v.* Baxendale (1854) 9 Exch. 341
13. Victoria Laundry (Windsor) Ltd *v.* Newman Industries Ltd (1949) 2 KB 528
14. Planche *v.* Colburn (1831) 8 Bing 14
15. Stevenson, Jordan and Harrison *v.* Macdonald & Evans (1951) 68 R.P.C. 190
16. The Ready-mixed Concrete (South East) Ltd *v.* The Minister of Pensions and National Insurance (1968) 1 All ER 433
17. Systems Floors (UK) Ltd. *v.* Daniel (1982) ICR 54; (1981) IRLR 475
18. Lister *v.* Romford Ice and Cold Storage Co. Ltd (1957) AC 535
19. Hivac Ltd *v.* Park Royal Scientific Instruments Ltd (1946) 1 All ER 350
20. Malik *v.* BCCI (1997) 3 All ER 1
21. Rowland *v.* Divall (1923) 2 KB 500
22. Beale *v.* Taylor (1967) 3 All ER 253
23. Wilson *v.* Rickett, Cockerell & Co. Ltd (1954) 1 All ER 868
24. Henry Kendall & Sons *v.* William Lillico & Sons Ltd (1968) 2 All ER 444
25. Ashington Piggeries Ltd *v.* Christopher Hill Ltd (1971) 1 All ER 847
26. George Mitchell (Chesterhall) Ltd. *v.* Finney Lock Seeds Ltd. (1983) 2 All ER 737
27. Ashdown *v.* Samuel Williams & Sons (1957) 1 All ER 35
28. Wheat *v.* Lacon & Co. Ltd (1966) 1 All ER 582

List of the acts referred to

Employment of Children Act 1973
The Misrepresentation Act 1967
Contract of Employment Act 1963
Equal Pay Act 1970
Sex Discrimination Act 1975
Race Relations Act 1976
Employment Protection (Consolidation) Act 1978
Sale of Goods Act 1893
Sale of Goods Act 1979
Unfair Contract Terms Act 1977
Occupier's Liability Acts 1957 and 1984
Health and Safety at Work etc. Act 1974
Health and Safety (Leasing Arrangements) Regulations 1980, SI 1980 No. 907
Supply of Goods and Services Act 1982

Consumer Protection Act 1987
Equal Pay (Amendment) Regulations 1983
Social Security Act 1986
Trade Union Reform and Employment Rights Act 1993
Unfair Terms in Consumer Contracts Regulations 1994, SI 1994 No. 3159
Employment Rights Act 1996

Chapter 5
Industrial relations law
R. D. Miskin

5.1 Introduction

Up until the end of the 1960s an employee in the UK had little or no legal protection so far as his employment was concerned. The employer had no duty to give the employee any specific form of contract or inform him of the basic terms of his employment. Thus, the employer had the power to dismiss virtually as and when he wanted and had no duty to give the employee the reasons for such dismissal, so the average employee, with only very few exceptions, had no continuity of employment nor any right to claim compensation when unfairly dismissed.

However, towards the end of the 1960s the then Government decided that an employee was entitled to be legally protected in the continuity of his employment, not to be unfairly or unreasonably dismissed and to be informed of the more important terms of his employment. There followed a number of Acts of Parliament implementing these rights and many others. It is the object of this chapter to consider the basic principles of such legislation and the rules, regulations and decided cases supporting it, but it must be appreciated that only an overall summary can be given. Where specific problems arise reference must be made to the relevant statutes.

5.2 Employment law

Employment law is governed by statute and the decided case law arising from those statutes. The most important is the Employment Rights Act 1996 (ERA) which consolidates earlier enactments relating to employment rights. It supersedes the Employment Protection (Consolidation) Act 1978, the Wages Act 1986 and parts of the Trade Union and Employment Rights Act 1993, many sections of which remain. However, it does not repeal the Equal Pay Act 1970, the Sex Discrimination Act 1975, the Race Relations Act 1976, the Transfer of Undertakings (Protection of Employment) Regulations 1981 or the Disability Discrimination Act 1995 but is the principal enactment to be taken into account when considering the rights of employees.

5.2.1 The Employment Rights Act 1996

This is 'An Act to consolidate enactments relating to employment rights' and deals mainly with the following areas:

(a) An employee's right to a written statement of the main particulars of his employment.
(b) Protection of wages.
(c) Guaranteed payments.
(d) Protection for employees working on a Sunday.
(e) The right not to suffer detriment in employment.
(f) Time off work.
(g) Suspension from work on, *inter alia*, health and safety grounds.
(h) Maternity rights.
(i) Termination of employment.
(j) The right not to be unfairly dismissed and remedies for unfair dismissal.
(k) Entitlement to redundancy payment.
(l) Position where the employer is insolvent.
(m) Definition of what amounts to a week's pay.

One of the main rights granted to an employee under this Act is that the employer must give a written statement of the particulars of his employment. Although it is often assumed that the written particulars form his 'contract' of employment this is not technically so. However, they amount to very strong evidence of the terms of a contract of employment and, from the purely practical point of view, they are the only so-called contract many employees receive. To prove that the terms are not of a contractual nature would be difficult.

The information must be given not later than two months after the commencement of employment and contain the following information:

(i) The names of both employer and employee.
(ii) The date when employment began.
(iii) The date on which the employee's period of continuous employment began.
(iv) The scale or rate of remuneration or the method of calculating it, whether such remuneration is paid weekly, monthly or at some other specified interval.
(v) Terms relating to hours of work.
(vi) Any entitlement to holiday, including public holiday and holiday pay.
(vii) Payment during incapacity for work due to sickness or injury including sick pay provisions.
(viii) Pensions and pension schemes.
(ix) Length of notice to which an employee is entitled and is required to give to terminate his contract of employment.
(x) Job title and, where the employment is not permanent, the period for which it is expected to continue.

(xi) Place of work.
(xii) Any collective agreements which affect the terms and conditions of employment.
(xiii) Details of necessity to work outside the UK if relevant.
(xiv) Details of any disciplinary rules that apply to the employee.
(xv) Name of person to whom complaints may be made if the employee is dissatisfied.

It should be noted that the employer may, so far as sickness, injury, pension schemes and collective agreements are concerned, specify the document or agreement in which the provisions are contained provided the employee has reasonable access to them to acquaint himself with their contents. An employee is entitled to an itemised pay statement and the employer should make no deductions from his wages unless the employee has specifically agreed to such deductions in writing. The exceptions to this rule are statutory deductions for income tax, National Insurance contributions and payments made under an Attachment of Earnings Order. This provides important protection for the employee and is firmly enforced by the Tribunals. Industrial Tribunals procedures and practices are set out in the Industrial Tribunals Act 1996 and associated regulations. Other provisions of ERA are considered in the following sections.

5.3 Discrimination

5.3.1 Sex discrimination

5.3.1.1 The Sex Discrimination Act 1975

Section 1 states that a person discriminates against a woman if in any circumstances relevant to the purposes of any provision of the Act he:

(a) on the ground of her sex, treats her less favourably that he treats or would treat a man, or
(b) applies to her a requirement or condition which applies or would apply equally to a man but:
    (i) which is such that the proportion of women who can comply with it is considerably less than the proportion of men who can comply with it, and
    (ii) which he cannot show to be justifiable, irrespective of the sex of the person to whom it applied, and
    (iii) which is to her detriment because she cannot comply with it.

Section 1(1)(a) is relatively straightforward in that a woman is discriminated against if she is treated less favourably than a man. This is known as direct discrimination and arose in *James* v. *Eastleigh Borough Council*[1]. It should be noted that the complainant does not have to prove that the discrimination was intentional, only that it occurred.

The circumstances covered by s. 1(1)(b) are known as indirect discrimination and require that the employer has to prove his conduct was justifiable. In *Home Office* v. *Holmes*[2] the Tribunal found, on the particular facts of the case, that the employer's action was not justifiable despite the fact that he produced detailed statistical and other evidence to support his claim. This case established that employers must accept the need to adjust working patterns to the needs of returning mothers.

Section 2 requires that men should receive equal treatment, but the vast majority of claims are from women. Section 6 concerns employment opportunities and says:

1 It is unlawful for a person to discriminate against a woman:
   (a) in the arrangements he makes for the purposes of determining who should be offered that employment, or
   (b) in terms on which he offers her that employment, or
   (c) by refusing, or deliberately omitting to offer her that employment.
2 It is unlawful for a person to discriminate against a woman employed by him:
   (a) in the way he affords her access to opportunities for promotion, transfer, training or any other benefits, facilities or services, or by refusing or deliberately omitting to afford her access to them, or
   (b) by dismissing her or subjecting her to other detrimental treatment.

Other parts of this section include further protection to a woman in employment.

There are exceptions where sex is a genuine occupational qualification and this is covered in s. 7 which provides:

1 In relation to sex discrimination:
   (a) s. 6(1)(a) or (c) do not apply to any employment where being a man is a genuine occupational qualification for the job, and
   (b) s. 6(2)(a) does not apply to opportunities for promotion or transfer to, or training for such employment.
2 Being a man is a genuine occupational qualification for a job only where:
   (a) the essential nature of the job calls for a man for reasons of physiology (excluding physical strength or stamina) or, in dramatic performances or other entertainment, for reasons of authenticity so that the essential nature of the job would be materially different if carried out by a woman; or
   (b) where the job needs to be held by a man to preserve decency or privacy because:
      (i) it is likely to involve physical contact with men in circumstances where they might reasonably object to it being carried out by a woman or
      (ii) the holder of the job is likely to do his work in circumstances where men might reasonably object to the presence of a

woman because they are in a state of undress or are using sanitary facilities or

(ba) the job is likely to involve the holder doing his work or living in a private home and need to be held by a man because objection might reasonably be taken to allowing a woman:
  (i) the degree of physical or social contact with a person living in the home or
  (ii) the knowledge of intimate details of such a person's life or of the home.

These are the main exceptions to the general rule but it should be noted that there are others which apply.

In health and safety matters discrimination is allowed on health grounds and concerning both pregnancy and maternity. This occurred in *Page* v. *Freight Hire Tank Haulage Ltd*[3] where a woman lorry driver, who was of child bearing age, was prevented from driving a tanker lorry containing chemicals that could be harmful to a woman's ability to bear children.

5.3.1.2 Sex Discrimination Act 1986

This Act amends certain provisions of the 1975 Act and in particular makes reference to collective agreements, partnerships, employment in private households, rules of professional bodies or organisations, exemptions for small businesses and discrimination in training

5.3.2 The Race Relations Act 1976

This Act is couched in almost exactly the same terms as the Sex Discrimination Act 1975 in that it provides in s. 1(1) that a person discriminates against another if in any circumstances relevant to the purposes of any provision he:

(a) on racial grounds treats that person less favourably than he treats or would treat other persons, or
(b) he applies to that person a requirement or condition which he applies or would apply equally to persons not of the same racial groups as that person but:
  (i) which is such that the proportion of persons of the same racial group as that person who can comply with it is considerably smaller than the proportion of persons not of that racial group who can comply with it, and
  (ii) which he cannot show to be justifiable irrespective of the colour, race, nationality or ethnic or national origins of the person to whom it is applied, and
  (iii) which is to the detriment of that other person because he cannot comply with it.

Race discrimination can be both direct and indirect, the latter being more difficult to recognise. Section 4 specifies that:

1. It is unlawful for a person, in relation to employment by him at an establishment in Great Britain, to discriminate against another:
    (a) in the arrangement he makes for the purpose of determining who should be offered that employment; or
    (b) in the terms on which he offers him that employment; or
    (c) by dismissing him or subjecting him to any other detrimental treatment.

There are exceptions where genuine occupational qualifications are required and these are listed in s. 5 as:

1. In relation to racial discrimination:
    (a) s. 4(1)(a) or (c) do not apply to any employment where being of a particular racial group is a genuine occupational qualification for the job and
    (b) s. 4(2)(b) does not apply to opportunities for promotion or transfer to, or training for, such employment.
2. Being of a particular racial group is a genuine occupational qualification for a job only when:
    (a) the job involves participation in a dramatic performance, or other entertainment in a capacity for which a person of that racial group is required by reasons of authenticity; or
    (b) the job involves participation as an artiste or photographic model in the production of a work of art, visual image or sequence of visual images for which a person of that racial group is required for reasons of authenticity; or
    (c) the job involves working in a place where food or drink is (for payment or not) provided to and consumed by members of the public or a section of the public in a particular setting for which a person of that racial group is required for authenticity; or
    (d) the holder of the job provides persons of that racial group with a personal service promoting their welfare, and those services can most effectively be provided by a person of that racial group.

There are restrictions on advertisements which might reasonably be understood to indicate, or do indicate, an intention to racially discriminate.

5.3.3 The Disability Discrimination Act 1995

This Act makes it unlawful to discriminate against any disabled person in connection with employment, the provision of goods, facilities and services or the disposal or management of premises, makes provision for the employment of disabled persons and establishes a National Disability Council. The Act defines disability and disabled persons and in s. 4 makes it unlawful for an employer to discriminate against a disabled person:

(a) in the arrangements which he makes for the purpose of determining to whom he should offer employment;
(b) in the terms on which he offers that person employment; or
(c) by refusing or offer to deliberately not offering him employment.

It is also unlawful for an employer to discriminate against a disabled person whom he employs:
    (i) in the terms of employment;
    (ii) in the opportunities afforded for promotion, transfer, training or receiving any other benefit;
    (iii) by refusing to offer, or deliberately withholding, any such opportunity; or
    (iv) by dismissing him or subjecting him to detrimental treatment.

Section 5(1) states, *inter alia*, that the employer discriminates against a disabled person if:

(a) for a reason which relates to the disabled person's disability he is treated less favourably than others to whom that reason does not, or would not, apply; and
(b) he cannot show that the treatment in question is justified,
    (aa) he fails to comply with a duty under s. 6 imposed on him in relation to the disabled person; and
    (bb) he cannot show that this failure to comply with that duty is justified.

Section 6 deals with the duty of the employer to make arrangements to enable a disabled person to carry out his job properly. The provisions of s. 6 have important health and safety connotations and should be studied in conjunction with the Disability Discrimination (Employment) Regulations 1996 and the Code of Good Practice on the Employment of Disabled People[4] which deals, in ss. 4.2, 4.3 and 4.4, with safety aspects in particular.

Note should also be taken of the Equal Pay Act 1970 which is aimed at preventing discrimination in terms and conditions of employment between men and women. It is a complex Act to understand but its main principle is to ensure that where a woman is employed on like work with a man on the same employment she is entitled to the same terms of employment as a man. The most obvious claim by a woman under this Act is that she should be paid at the same rate as a man. Although the procedures for bringing such a claim are complex, there have been many Industrial Tribunal cases where this particular point has been argued.

5.3.4 The Health and Safety (Young Persons) Regulations 1997

These Regulations modify the Management Regulations (MHSW) with respect to the employment of young person and in reg. 2 give two important definitions; firstly, that a child is a person who is not over compulsory school age, and, secondly, that a young person is one who has not attained the age of 18.

An employer must not employ a young person unless he has carried out a risk assessment taking particular account of the following factors:

(a) the inexperience, lack of awareness of risks and immaturity of young persons;
(b) the fitting out and layout of the workplace and the work station;
(c) the nature, degree and duration of exposure to physical, biological and chemical agents;
(d) the form, range and use of work equipment and the way in which it is handled;
(e) the organisation of processes and activities;
(f) the extent of the health and safety training provided for young persons; and
(g) the risks from agents, processes and work listed in the annex to EU Directive 94/33/EC on the Protection of Young People at Work[5].

Prior to employing a child, an employer must provide a parent of the child with comprehensible and relevant information on:

(a) the risks to his health and safety identified by the assessment;
(b) the preventive and protective measures; and
(c) the risks notified to him by other employers sharing the same premises.

An employer must not employ a young person for work:

(a) which is beyond his physical and psychological capacity;
(b) involving harmful exposure to agents which are toxic, carcinogenic, cause heritable genetic damage, harm to the unborn child or which in any way chronically affect human health;
(c) involves harmful exposure to radiation;
(d) involves the risk of accidents which it may reasonably be assumed cannot be recognised or avoided by young persons owing to their insufficient attention to safety or lack of experience or training; or
(e) in which there is a risk to health from:
    (i) extreme cold or heat
    (ii) noise or
    (iii) vibration.

A young person, who is no longer a child, may be employed for work:

(a) where it is necessary for his training
(b) which is supervised by a competent person and
(c) where any risks have been reduced to the lowest level that is reasonably practicable.

5.3.5 Joint consultation

In October 1978, the Safety Representatives and Safety Committee Regulations 1977 (SRSC) came into effect and gave to those unions that were recognised in the workplace the right to appoint safety representatives. Those safety representatives were given certain functions and employers were required to give to the representatives, to enable them to perform their functions, time off work with pay for training and to carry out their functions, information necessary to fulfil their functions and allow them to carry out inspections of the workplace following accidents.

In 1989 the European Council adopted a directive no. 89/391/EEC (known as the Framework Directive) which contained a requirement for workpeople, whether union members or not, to be consulted about matters concerning their health and safety at work. The UK Government held that this was covered by the SRSC but a judgement by the European Court of Justice[6] in 1992 established the right of all employees to be consulted. This right was brought into effect in the UK by the Health and Safety (Consultation with Employees) Regulations 1996.

These two Regulations effectively give the same rights and functions to safety representatives, whether union or employer appointed, and include the right:

1. to be consulted on:
   (a) the introduction of measures affecting health and safety
   (b) arrangements for appointing safety advisers
   (c) arrangements for appointing fire and emergency wardens
   (d) the health and safety information to be provided to employees
   (e) provisions for health and safety training
   (f) health and safety implications of new technologies
2. to be given sufficient information:
   (a) to carry out their functions
   (b) on accidents that had occurred but not:
      - where an individual can be recognised
      - if it could prejudice the company's trading
      - on matters subject to litigation
      - if it was against national security
      - if it contravened a prohibition imposed by law
3. to have time off work with pay:
   - to carry out their functions
   - to receive training
4. to carry out their functions which include:
   - making representations to the employer on hazards and incidents affecting his constituents' health and safety
   - being the contact with and receive information from HSE inspectors
   - investigating potential hazards and incidents affecting those he represents.

In addition, union appointed safety representatives have the right to:

- investigate complaints by those they represent
- carry out inspections of the workplace subject to the agreement of the employer
- attend meetings of the safety committee.

In both cases, complaints against the employer concerning refusal to provide, or allow time off for, training and for not paying for that time off are heard by an Industrial Tribunal.

5.3.6 Working time

The Working Time Regulations 1998, which derive from EU directive no. 93/104/EC[7], provide for a maximum working week of 48 hours. However, this can be extended by written agreement between the employee and employer. Night work is restricted to 8 hours in each 24 hour period and night workers are to have health assessments. Employers are required to keep records of the hours worked. Periodic rest times are specified as are the rest breaks to be taken if the working period is more than 6 hours ($4\frac{1}{2}$ hours for young persons). Workers (except those in agriculture) are entitled to 4 weeks paid leave each year.

Excluded from these requirements are workers in transport, trainee doctors, sea fishermen, police and armed forces and domestic servants. The Regulations allow the employer to vary working times to meet particular employment and trading circumstances.

5.4 Disciplinary procedures

Dismissal is dealt with in the following section, but a Tribunal, to find a dismissal fair, must be satisfied that the dismissal was reasonable in all the circumstances. In the majority of cases this entails the employer following his own disciplinary and grievance procedures. It is important that an employer should have formal disciplinary rules which should be communicated to each and every employee. Should such communication not have taken place the employer will not be able to rely on such rules, and a dismissal which might otherwise have been fair could be ruled unfair.

Acceptable procedures are outlined in a Code of Practice[8] and referred to in an ACAS advisory handbook[9] which emphasise the importance of such rules. The Code of Practice states in paragraph 6 that:

> 'it is unlikely that any set of disciplinary rules can cover all circumstances that may arise: moreover, the rules required will vary according to the circumstances, such as the type of work, working conditions and size of the establishment. When

drawing up rules the aim should be to specify clearly and concisely those necessary for the efficient and safe performance of the work and for the maintenance of satisfactory relations within the workforce and between employees and management. Rules should not be so general as to be meaningless.'

Communicating the procedures to employees is dealt with in paragraph 7 which states that:

'Rules should be readily available and management should make every effort to ensure that employees know and understand them. This may be best achieved by giving every employee a copy of the rules and by explaining them orally. In the case of new employees, this should form part of an induction programme.'

To be effective the procedures should:

(a) be in writing,
(b) specify to whom they apply,
(c) provide for matters to be dealt with quickly,
(d) indicate the disciplinary actions that can be taken,
(e) specify the various levels in the organisation that have the authority to take various forms of disciplinary action, ensuring that the immediate supervisor's powers to dismiss are restricted and require reference to a senior manager,
(f) provide for individuals to be informed of the complaints against them and be given an opportunity to state their case before decisions are reached,
(g) give individuals the right to be accompanied by a trade union representative or by a fellow employee of their choice,
(h) ensure that, except for gross misconduct, no employee is dismissed for a first breach of discipline,
(i) ensure that, disciplinary action is not taken until the case has been carefully investigated,
(j) ensure that the individual is given an explanation for any penalty imposed, and
(k) provide a right of appeal and specify the procedure to be followed.

A record should be kept of any disciplinary actions taken against an employee for breach of the rules including lack of capability, conduct etc. and what disciplinary action was taken and the reasons supporting such action. The disciplinary procedures should be reviewed from time to time to ensure that they comply with the then practices of the employer. A written record should be kept of an oral warning to prove that it was actually given.

Many of these rules and procedures will incorporate items relevant to safety, health and welfare of the employees in that particular employment. The emphasis placed on particular aspects of safety and health will reflect the degree of risk or hazard faced by the employee in his daily work and what effect failure to follow these rules might have on the employees themselves, the environment or the continuing operation of the business. The onus is on the employer to draw up these rules and he may do this unilaterally but it is more prudent of him to consult the employees or their representative to obtain agreement to and acceptance of the various procedures before they are implemented.

The employer should ensure that, except for gross misconduct, no employee is dismissed for a first breach of discipline. Instead the employer should operate a system of warnings consisting of an oral warning, a first written warning and then a final written warning before dismissal is considered.

An employee at any disciplinary hearing must be informed of his right to appeal.

Appendix 3 of the ACAS advisory handbook[9] includes examples of disciplinary procedures which cover the appropriate points.

Until the case of *Polkey* v. *A.E. Dayton (Services) Ltd*[10] the courts tended to take the view that where employers did not follow their disciplinary procedures, but even if they had it would have made no difference to the outcome, then the dismissal was fair notwithstanding such failure. This principle was summarised by Browne Wilkinson, J. in *Sillifant* v. *Powell Duffryn Timber Ltd*[11] as follows:

> 'Even if, judged in the light of circumstances known at the time of dismissal, the employer's decision was not reasonable because of some failure to follow a fair procedure, yet the dismissal can be held to be fair if, on the facts proved before the Industrial Tribunal, the Industrial Tribunal comes to the conclusion that the employer could reasonably have decided to dismiss if he had followed fair procedure.'

The *Polkey* decision found that the one question an Industrial Tribunal was not permitted to ask in applying the test of reasonableness was the hypothetical question of whether it would have made any difference to the outcome if the appropriate procedural steps had been taken. However, it was quite a different matter if the Tribunal was able to conclude that the employer himself, at the time of dismissal, acted reasonably in taking the view that, in the exceptional circumstances of the particular case, the procedural steps normally appropriate would have been futile, could not have altered the decision to dismiss and therefore could be dispensed with. In such a case the test of reasonableness may have been satisfied.

The *Polkey* decision makes it clear that a Tribunal will not err in law if it starts from the premise that breach of procedures, at least where they embody significant safeguards for the employee, will render a dismissal unfair. It is important that the employer follows his disciplinary procedures as closely as possible in the circumstances of any particular case.

5.5 Dismissal

Under s. 94 of ERA an employee has the right not to be unfairly dismissed by his employer but s. 95(1) allows dismissal if:

(a) the contract under which he is employed is terminated by the employer (whether with or without notice);
(b) he is employed under a contract for a fixed term and that term expires without being renewed under the same contract; or
(c) the employee terminates the contract under which he is employed (with or without notice) in circumstances in which he is entitled to terminate it without notice by reason of the employer's conduct.

Also, in s. 95(2) an employee shall be taken to be dismissed by his employer if:

(a) the employer gives notice to the employee to terminate his contract of employment; and
(b) at a time within the period of that notice the employee gives notice to the employer to terminate the contract on a date earlier than the date on which the employer's notice is due to expire and the reason for the dismissal is taken to be the reason for which the employer's notice is given.

In s. 97 'the effective date of termination' of employment is taken as:

(a) the date on which the notice expires whether the notice is given by the employee or the employer;
(b) the date on which termination takes effect if terminated without notice; and
(c) the date of expiry of the contract where it is a fixed term and is not being renewed.

An employee who has had his employment terminated is entitled to written reasons from the employer.

In determining whether the dismissal is fair or unfair, s. 98(1) requires the employer to show:

(a) the reason (or if more than one the principal reason) for the dismissal, and
(b) that it is either a reason falling within s. 98(2) or some other substantial reason sufficient to justify the dismissal of an employee from the position which he held.

A reason for dismissal is sufficient if:

(a) it relates to the capability or qualifications of the employee to perform work of the kind he is employed to do,
(b) it relates to the conduct of the employee,
(c) the employee is redundant, or

(d) the employee could not continue to work in the position which he held without contravention (either on his part or on the part of the employer) of a statutory duty or restriction.

The above are reasons upon which an employee can be fairly dismissed; however, s. 98(4) states that where an employer has fulfilled the requirements of subs. (1), the determination of the question whether the dismissal is fair or unfair:

(a) depends on whether, in the circumstances (including the size and administrative resources of the employer's undertaking), the employer acted reasonably or unreasonably in treating the reason as being sufficient for dismissing the employee, and
(b) shall be determined in accordance with equity and the substantial merits of the case.

It is important to understand the grounds upon which an employer can rely as having acted reasonably and fairly in the dismissal of an employee. The above reasons for dismissal are considered in the following sections.

5.5.1 Capability or qualification

Capability is defined in ERA as the employee's capability assessed by reference to skill, attitude, health or other physical or mental quality and such qualifications as any degree, diploma or other academic, technical or professional qualification relevant to the position the employee held. The two main classes of capability, or lack of it, are ill-health and the inability of the employee to carry out his duties in a reasonable and acceptable manner.

5.5.2 Ill-health

Ill-health falls into two categories. Firstly, where the employee is sick or incapacitated for one long period and, secondly, where he has regular short spells of illness which, added together, represent a lengthy period of absence. It is necessary to consider these two classes of illness separately as the legal position is different in each case.

5.5.2.1 Long-term illness

The leading case which sets out the main principles to support a fair dismissal for long-term illness is *Spencer* v. *Paragon Wallpapers Ltd*[12] in which the employee had been absent sick for approximately two months and the medical opinion was that he would return within another four to six weeks. The EAT held that in such cases the employer must take into account:

(a) the nature of the illness,
(b) the likely length of the continuing absence,
(c) the employer's need for the work to be done, and
(d) the availability of alternative employment for the employee.

Since all four criteria had been met, the dismissal was fair.

Consultation with the employee and investigation of the medical position by the employer would seem to be the two most important criteria. In *East Lindsay District Council* v. *Daubny*[13] the EAT stated that unless there were wholly exceptional circumstances the employee should be consulted and the matter discussed with him before his employment was terminated on the ground of ill-health.

5.5.2.2 Continuing periodic absences

In stark contrast to the above there have been several EAT cases where the employee has been dismissed for persistent absenteeism due to a succession of short illnesses. In *International Sports Company Ltd* v. *Thomson*[14] an employee was persistently absent for minor ailments that could not later be confirmed by medical examination. After review by the employer of her absence record and being given reasonable warnings she was dismissed. There had been no improvement in her attendance and the dismissal was held to be fair.

A factor in this decision was that there had been no medical investigation and that the employer would have been no wiser even if he had carried out such examination.

Further, it is essential for the employer to stick to his disciplinary procedures and give the appropriate warnings.

Ill-health caused by an employee's duties can lead to fair dismissal but an employer may be held not to have acted reasonably if reasonable steps had not been taken to eliminate the danger to health stemming from the job. If an employee claims he is absent for reasons of ill-health but the employer believes he is malingering it may be difficult for the employer to prove such. However, in *Hutchinson* v. *Enfield Rolling Mills Ltd*[15] the Tribunal was satisfied the employer had done so. They held that if there is evidence to suggest that the employee is, in fact, fit to work, despite his having a doctor's sick note, the employer can seek to rely upon that evidence to justify the dismissal.

5.5.3 Lack of skill on the part of the employee

Tribunals often find it difficult to decide whether a dismissal for incompetence is fair or unfair. What is clear is that it is not open to them to rely on their own view as to the employee's competence rather than that of the employer's. In *Taylor* v. *Alidair Ltd*[16] Lord Denning made this clear when he said 'Whenever a man is dismissed for incapacity or incompetence it is sufficient that the employer honestly believes on reasonable grounds that the man is incapable and incompetent. It is not necessary for the employer to prove that he is in fact incapable or incompetent.'

The test therefore is the genuine belief of the employer based on the evidence that he has gathered to show that his view is a reasonable one. In such cases the employer must rely on his disciplinary procedures and give constructive warnings to the employee to give him an opportunity of improving his performance. Thus the employer should carry out a thorough evaluation of the employee's performance and discuss his criticisms with the employee personally, warn the employee of the consequence of there being no improvement and then give him reasonable opportunity to improve.

There are cases where an employer cannot follow the above procedure because of the seriousness of the consequences of further error. In *Taylor* v. *Alidair Ltd* a pilot was dismissed and not given any further opportunities to improve when the company considered he was to blame when he made a faulty landing which caused considerable damage to the aircraft. In this case, the Court of Appeal specifically approved of the following statement made by Bristow, J.: 'In our judgement there are activities in which the degree of professional skill which must be required is so high, and the potential consequences of the smallest departure from that high standard are so serious, that one failure to perform in accordance with those standards is enough to justify dismissal. The passenger-carrying airline pilot, the scientist operating the nuclear reactor, the chemist in charge of research into the possible effects of, for example, thalidomide, the driver of the Manchester to London express, the driver of the articulated lorry full of sulphuric acid, are all in a position in which one failure to maintain the proper standard of professional skill can bring about a major disaster.'

Finally, the employer is entitled to dismiss an employee without warning where there is little likelihood of the employee improving his performance and his continuing presence is prejudical to the company's best interest. This is illustrated by the case of *James* v. *Waltham Holy Cross UDC*[17].

5.5.4 Misconduct

Misconduct in the place of work, or in certain circumstances outside it, is one of the major reasons for dismissal of an employee. It was defined by the Scottish EAT as 'Actings [sic] of such a nature, whether done in the course of employment or out of it, that reflect in some way on the employer–employee relationship.'

Discipline for misconduct falls into two main categories: firstly, the lesser transgressions should be dealt with under the employer's disciplinary practices, by way of warning and encouragement not to transgress again and, secondly, the more serious cases of gross misconduct, by instant dismissal. The employer should list in his disciplinary rules those acts that fall into the category of gross misconduct so that the employee is in no doubt whatsoever that by committing such act he renders himself liable to instant dismissal. The acts concerned vary from business to business but normally include:

- theft
- fraud
- deliberate falsification of records
- fighting
- assault on another person
- deliberate damage to company property
- serious incapability through alcohol or being under the influence of illegal drugs
- serious negligence which causes unacceptable loss, damage or injury
- serious acts of insubordination.

As well as referring to his disciplinary rules and procedures, an employer should refer to the contract of employment to ascertain what was required of the employee.

So far as criminal offences committed away from the place of work are concerned, the Code of Practice[8] makes it clear that these should not constitute automatic reasons for dismissal, but should be considered in the light of whether the offence in question makes the employee unsuitable for his or her type of work or unacceptable to other employees. If an employee secures employment by not disclosing a previous criminal conviction his dismissal on that ground is often fair provided he is not permitted to withhold such conviction under the Rehabilitation of Offenders Act 1974.

In cases of dismissal for misconduct it is essential that the employer has acted reasonably and fairly in all the circumstances. Although decided in 1978, the case of *British Home Stores* v. *Burchell*[18] still provides the basic guidelines to whether or not an employer has acted reasonably. The judgement in that case clearly sets out the steps an employer must take before dismissing an employee on the grounds of gross misconduct as:

1 belief in the employee's guilt,
2 having reasonable grounds for believing so, and
3 having carried out reasonable investigation to verify the grounds for sustaining that belief.

If the employer has followed these steps, then the Tribunal must uphold his decision although they may not necessarily have come to the same view themselves. Further, the standard of proof which an employer must meet is only that he should be satisfied on the balance of probabilities, and not beyond all reasonable doubt. These principles have been slightly eroded by subsequent legislation in that it is not essential for the last two elements to be proved but it will be very much in the employer's favour if he can do so.

There are a number of cases relating to misconduct which revolve around issues of health and safety in the workplace. The first of these is *Austin* v. *British Aircraft Corporation Ltd*[19] where the employer's attitude was considered unreasonable. Mrs Austin and her fellow employees were required to wear eye protection. Mrs Austin already wore glasses and the goggles provided were uncomfortable. However, she persevered for three months but eventually stopped wearing them. She raised the problem with her employers and the matter was put in the hands of the safety officer. Six months later nothing had been done so Mrs Austin resigned. The Tribunal hearing her case concluded that Mrs Austin had been constructively dismissed and was entitled to resign by reason of her employer's conduct.

The same principle applied in *Keys* v. *Shoefayre Ltd*[20] where the owner of a retail shop failed to take proper security precautions to protect his employees who worked in a shop in an area with a high crime rate that had suffered two armed robberies. Here it was held that the employer had failed to take reasonable care and provide a safe system of work and that Mrs Keys' resignation amounted to unfair constructive dismissal.

In the manufacture of tyres, part of the process emits dust and fumes that reports from America indicated might be carcinogenic. Negotiations resulted in face masks being provided as an interim measure until expensive capital equipment could be obtained which would improve matters, a step that was supported by the HSE. However, in *Lindsay* v. *Dunlop Ltd*[21] the employee was not satisfied with these precautions and delayed removing the tyres from the press until the fumes had dispersed. This seriously affected production and, following discussion with his union, the employer dismissed the employee. The Tribunal held that the dismissal was fair, a decision upheld by the Court of Appeal on the grounds that the employer had taken all reasonable steps in the circumstances.

With regard to criminal offences committed outside employment, the ACAS code holds that these should not be treated as automatic reasons for dismissal and that the main consideration should be whether the offence renders the employee unsuitable for his or her job or unacceptable to other employees. Employees should not be dismissed solely because a charge against them is pending or because they are absent through having been remanded in custody.

5.5.5 Redundancy

The provisions in ERA regarding redundancy are both technical and difficult to understand but s. 139 states:

1 an employee who is dismissed shall be taken to be dismissed by reason of redundancy if the dismissal is wholly or mainly due to the fact that:
    (a) his employer has ceased or intends to cease:
        (i) to carry on the business for the purposes of which the employee was employed by him, or
        (ii) to carry on that business in the place where the employee was employed, or
    (b) the requirements of the business:
        (i) for the employee to carry out work of a particular kind, or
        (ii) for the employee to carry out work of a particular kind in the place where the employee was employed
    have ceased or diminished or are expected to cease or diminish.

2 For the purposes of subs. (1) the business of the employer together with the business or businesses of his associated employers shall be treated as one (unless either of the conditions specified in paragraphs (a) and (b) of that subsection would be satisfied without so treating them).

It is open to an employee to claim that the employer acted unreasonably in electing to make workers redundant. He may allege that his dismissal on this ground was unfair for two reasons. The first is that the method of selection was unfair and the second that within the meaning of s. 98(4)–(6) his selection was unreasonable. It is automatically unfair to select an employee for redundancy for:

1 a pregnancy or pregnancy related reason
2 a health and safety reason specified in s. 100
3 a reason related to the fact that they are protected shop workers specified in s. 101
4 being trustee of an occupational pension scheme specified in s. 102
5 being a representative or candidate to be such representatives as specified in s. 103
6 a reason relating to the assertion of a statutory right under s. 104
7 a reason connected with trade union membership or activities.

In *Williams* v. *Compair Maxam Ltd*[22] the EAT set out a general guideline as to the circumstances in which a selection for redundancy would be fair. These were that the employer should:

1 seek to give as much notice as possible;
2 consult the union as to the criteria to be applied in selecting the employees to be made redundant;
3 ensure that such criteria do not depend solely upon the opinion of the person making the selection but can be objectively checked against such things as attendance record, efficiency at the job, experience or length of service;
4 seek to ensure that the selection is made fairly in accordance with these criteria and consider any representation the union may make; and
5 see whether, instead of dismissing an employee, he could offer him alternative employment.

Whether or not a union is involved, a sensible employer will follow the above rules.

Over the years, the courts have varied in the importance they put upon consultation between the employer and the employee where redundancy is concerned. However, the present position is that consultation is of considerable importance and in *Polkey*[10] Lord Bridge said '... in the case of redundancy, the employer will normally not act reasonably unless he warns and consults any employee affected or their representative'. This does not mean that where no consultation takes place with the employee the redundancy is inevitably unfair but it certainly makes the employer's position more difficult to sustain. In some cases there is a statutory obligation to consult recognised trade unions over redundancies. A redundancy can also be rendered unfair by the failure of the employer to find alternative employment for the employee.

5.5.6 Contravention of an enactment

An employer can dismiss fairly where he can show that the employee could not continue to work without contravening a statutory enactment. The most common example is where an employee is disqualified from driving by a court and the principal part of his job is driving. In these circumstances the employer would have to consider either providing alternative transport or alternative employment. Other instances occur where an Authority is directed under the Schools Regulations not to employ a teacher because he is unsuitable or where an airline pilot cannot fly because an Air Navigation Order stipulates that he cannot do so unless the operator has satisfied himself that the pilot is competent to perform his duties.

5.5.7 Any other substantial reason

Cases arise where an employer cannot satisfy the Tribunal that a redundancy had arisen, but the Tribunal may allow a plea of 'any other substantial reason' instead. In *R.S. Components Ltd* v. *Irwin*[23] the employers manufactured electrical components and members of their staff left, set up in competition and solicited the former employer's customers. As the employers were losing money they sought to impose a restrictive covenant on their staff to prevent it happening again. Four employees refused to accept the new terms and were dismissed. The EAT held that the dismissal was fair as the employer was dismissing for any other substantial reason.

In *Sanders* v. *Scottish National Camps Association Ltd*[24] the employee was dismissed from his job as a maintenance handyman at a children's camp when it was discovered that he was homosexual and had been involved in a homosexual incident, though not involving children. In all the circumstances his dismissal was found to be fair. In the case of *Treganowan* v. *Robert Knee & Co. Ltd*[25] a dismissal was found to be fair where a personality difference had arisen between the employee and other girls in the office creating an intolerable atmosphere. Leaking confidential information has also been admitted as any other substantial reason.

5.5.8 Reasons for dismissal which are automatically unfair

Of particular importance to those involved in dismissals involving health and safety issues are subss. (1), (2) and (3) of s. 100 of ERA whereby:

1. An employee is regarded as having been unfairly dismissed if the principal reason for the dismissal was:
   (a) having been designated by the employer to carry out activities in connection with preventing or reducing risks to health and safety at work, the employee carried out, or proposed to carry out, such activities,
   (b) being a representative of workers on matters of health and safety at work or a member of a safety committee:
      (i) in accordance with arrangements established under or by virtue of any enactment,
      (ii) by reason of being acknowledged as such by the employer, the employee performed, or proposed to perform, the functions of a representative or a member of a safety committee,
   (c) being an employee at a workplace where:
      (i) there was no such representative or safety committee, or
      (ii) there was a representative or safety committee but it was not reasonably practicable for the employee to raise matters by those means, he brought to his employer's attention, by reasonable means, circumstances connected with his work which he reasonably believed were harmful or potentially harmful to health or safety,
   (d) in circumstances of danger which the employee reasonably believed to be serious and imminent and which he could not reasonably have been expected to avert, he left, or proposed to leave, or, while the danger persisted refused to return to, his place of work or any dangerous part of it, or
   (e) in circumstances of danger which the employee reasonably believed to be serious and imminent he took, or proposed to take, appropriate steps to protect himself and others from danger.
2. For the purposes of subs. (1)(e) whether steps which the employee took, or proposed to take, were appropriate is to be judged by reference to all the circumstances including, in particular, his knowledge and the facilities and advice available to him at the time.
3. Where the reason, or if more than one the principal reason, for the dismissal of an employee is that specified in subs. (1)(e), he shall not be regarded as unfairly dismissed if the employer shows that it was, or would have been, so negligent of the employee to take the steps which he took, or proposed to take, that a reasonable employer might have dismissed him for taking, or proposing to take, them.

The provisions of s. 100 create a number of difficult problems for the Tribunal in that they need to decide whether the employee used his rights reasonably or whether, deliberately or otherwise, he abused them. An important point is found in subs. (1)(e) where a union official might order individual employees to cease work until a potential hazard has been removed. The Act makes it clear that whether the steps taken by the official were appropriate or not must be determined by all the circumstances, including the facilities and advice available at the time.

Other examples of where an employee is regarded, prima facie, as having been unfairly dismissed include:

- pregnancy and connected reasons
- assertion of a statutory right
- shop workers and betting shop workers who refuse to work on Sundays
- being a trustee of an occupational pension scheme
- being an employees' representative
- matters relating to union membership or non-membership
- the transfer of an undertaking
- spent convictions.

5.5.9 Exclusions from the right to claim unfair dismissal

Instances where an employee is excluded from claiming unfair dismissal include:

1 persons who have not worked for the employer for the qualifying period of two years
2 persons beyond the relevant retiring age
3 persons ordinarily working abroad
4 in some cases, people dismissed in connection with a strike or lock-out
5 where the contract or terms of employment are illegal
6 share fishermen
7 persons employed in the Police Service
8 certain Crown employees.

These are the principal exclusions but there are others specified in the Act. The need for an employee to serve a two-year qualifying period to establish his right to be able to claim unfair dismissal is under review and may be reduced or dispensed with altogether.

5.5.10 Rights of an employee who has been unfairly dismissed

An employee who has been found to have been unfairly dismissed is entitled to either compensation or reinstatement and re-engagement, the latter applying where the original job from which he was unfairly dismissed is no longer available. Section 114 of ERA defines an order for reinstatement as 'an order that the employer shall treat the complainant in all respects as if he had not been dismissed'. It follows from this that his original terms of employment once more apply and he is entitled to the benefit of any improvement, such as an increase in pay that he would have had if he had not been unfairly dismissed.

When making an order for reinstatement or re-engagement, the Tribunal must consider whether or not it is practicable to do so. If the employer fails to comply with such an order then the Tribunal must again consider the question whether or not it was practicable for him to comply with it. Where such an order is not complied with and the employer cannot show that it was not practicable for him to comply with it, then an additional or penal compensation can be ordered against the employer.

The other award a Tribunal can make is that of compensation. Such an award is made up of two factors: firstly, the basic award is the equivalent of the statutory redundancy payment the employee would have received if he had been dismissed for that reason. The second is the compensatory award for the financial loss which the employee has suffered. It falls under various heads which include, *inter alia*, loss of salary, future loss of salary, estimated fluctuations in earnings, future loss of unemployment benefit and loss of pension rights. A percentage of the award can be ordered to be deducted by the Tribunal if they feel that the employee has contributed in any way to his own dismissal. It is the employee's duty to mitigate his loss and he must be able to satisfy the Tribunal that he has sought other employment but without success.

5.6 Summary

The main purpose of industrial relations legislation is to regulate the relationship between employer and employee and to determine the role and powers of trade union representatives in deciding the terms and conditions under which an employee has to work. It has become practice to include under the wing of 'industrial relations' anything that can affect the way in which an employee has to work, and in this respect safety has an important role to play.

This chapter has shown some of the ways in which decisions and actions taken for safety reasons can materially affect the employee's working conditions and, conversely, the ways in which employment legislation can affect safety issues. For the safety adviser to be able to perform his duties properly he must be aware of the wider implications of the recommendations he makes, particularly in the field of working conditions.

The law governing industrial relations is extremely complex and covers much more ground than it has been possible to cover in this chapter, but the most important of the statutory provisions have been covered briefly and some of the ways in which their application can affect the employer–employee relationship have been shown.

References

1. James *v.* Eastleigh Borough Council (1990) IRLR 288
2. Home Office *v.* Holmes (1984) IRLR 299; (1984) 3 All ER 549; (1985) 1 WLR 71
3. Page *v.* Freight Hire Tank Haulage Ltd (1980) ICR 29; (1981) IRLR 13
4. The National Disability Council, *Code of Good Practice on the Employment of Disabled People*, The Disability Council, London
5. EU Directive 94/33/EC *on the protection of young people at work*, EU, Luxembourg (1994)
6. European Court of Justice cases:
   C382/92, *Safeguarding of employee rights in the event of transfer of undertakings*, Celex no. 692JO382
   C383/92, *Collective redundancies*, Celex no. 692JO383, EU, Luxembourg (1992)

7. EU Directive 93/104/EC *concerning certain aspects of the organisation of working time* (The Working Time Directive), EU, Luxembourg (1993)
8. Advisory, Conciliation and Arbitration Service, Code of Practice No. 1, *Disciplinary practice and procedures in employment*, HMSO, London (1978)
9. Advisory, Conciliation and Arbitration Service, *The ACAS advisory handbook: Discipline at Work*, HMSO, London (1989)
10. Polkey *v.* A.E Dayton (Services) Ltd (1988) IRLR 503; (1987) All ER 974, HE (E)
11. Sullifant *v.* Powell Duffryn Timber Ltd (1983) IRLR 91
12. Spencer *v.* Paragon Wallpapers Ltd (1976) IRLR 373
13. East Lindsay District Council *v.* Daubney (1977) IRLR 181
14. International Sports Co. Ltd *v.* Thomson (1980) IRLR 340
15. Hutchinson *v.* Enfield Rolling Mills (1981) IRLR 318
16. Taylor *v.* Alidair Ltd (1978) IRLR 82
17. James *v.* Waltham Holy Cross UDC (1973) IRLR 202
18. British Home Stores *v.* Burchell (1978) IRLR 379
19. Austin *v.* British Aircraft Corporation Ltd (1978) IRLR 332
20. Keys *v.* Shoefayre (1978) IRLR 476
21. Lindsay *v.* Dunlop Ltd (1979) IRLR 93
22. Williams *v.* Compair Maxam Ltd (1982) ICR 800
23. R.S. Components *v.* Irwin (1973) IRLR 239
24. Sanders *v.* Scottish National Camp Association (1980) IRLR 174
25. Treganowan *v.* Robert Knee & Co. Ltd (1975) IRLR 247; (1975) ICR 405

Chapter 6

Consumer protection

R. G. Lawson

The expression 'consumer protection', and with it the notion of 'consumer law' first found expression in the Final Report of the Committee on Consumer Protection (Cmnd 1781) 1962[1], which led to the enactment of the Trade Descriptions Act in 1968[2]. This latter can fairly be regarded as the starting point of the modern law of consumer protection.

In more recent years much of the impetus for new consumer legislation has come from the UK's membership of the European Union (EU) as witnessed by the enactment in the UK of the Control of Misleading Advertisements Regulations 1988[3], the Consumer Protection Act 1987[4], the General Product Safety Regulations 1994[5] and the Unfair Terms in Consumer Contracts Regulations 1994[6], in each case derived from an EU directive. A further directive on comparative advertising was adopted by the EU Council in 1997 and currently awaits implementation in the UK. Also, discussions are now at an advanced stage on a directive on guarantees and consumer redress in relation to contracts of sale.

6.1 Fair conditions of contract

It is central to any system of consumer protection that a potential customer is given only truthful and accurate information about the goods and services that he is wanting to buy. Even before Parliament had decided to intervene, the courts had already decided to allow a remedy where a contract had been induced by fraud or misrepresentation. Where a consumer has been duped into entering a contract through deception of the kind practised by some salesmen, he would be given the right to put an end to the contract and claim compensation for any loss which he may have suffered. This development in the courts was eventually confirmed by the Misrepresentation Act 1967[7].

Valuable though these controls were, they applied only to what is called the civil law, i.e. the law which regulates the relations between citizens. Where a consumer had been the victim of fraud or misrepresentation, the initiative lay entirely upon him to take remedial action. It was only with the advent of the Trade Descriptions Act that the criminal law came to the aid of the consumer in such cases.

6.1.1 False trade descriptions

The main feature of the Trade Descriptions Act 1968[2], is to penalise the use of false trade descriptions. Section 2 of the Act contains an exhaustive list of what constitutes a false trade description for the purposes of the Act. Anything not in the list is not a trade description for the purposes of the Act. The list includes any statement as to:

(a) quantity, size or gauge;
(b) method of manufacture, production, processing or reconditioning;
(c) composition;
(d) fitness for purpose, strength, performance, behaviour or accuracy;
(e) any physical characteristics not included in the preceding paragraphs;
(f) testing by any person and results thereof;
(g) approval by any person or conformity with a type approved by any person;
(h) place or date of manufacture, production, processing or reconditioning;
(i) person by whom manufactured, produced, processed or reconditioned;
(j) other history, including previous ownership or use.

Included within this list shall be matters concerning:
(i) any animal, its sex, breed or cross, fertility and soundness;
(ii) any semen, the identity and characteristics of the animal from which it was taken and measure of dilution.

Quantity is defined to include length, width, height, area, volume, capacity, weight and number.

This list can be summarised as saying that any statement about goods which when made can be either true or false is a trade description. This has meant that a statement made in relation to a bar of chocolate that it was of 'extra value' was not a trade description since it was such a vague kind of claim that no one could say of it that it was true or false: *Cadbury Ltd* v. *Halliday*[8].

While it is true to say that this part of the Act seems to have been used almost exclusively to control some of the more dubious antics of the second-hand car trade (convictions for turning back a mileometer have been particularly common) this is far from being entirely the case. One good example arose in the case of *British Gas Board* v. *Lubbock* (1974)[9]. A gas cooker was advertised as being ignited by a hand-held ignition pack. At the time the advertisement was shown, this was no longer true. The Board was prosecuted and convicted for making a false statement about the composition and the physical characteristics of goods.

Another example is the decision in *Queensway Discount Warehouses* v. *Burke* (1985)[10]. A wall unit was advertised in the national press. It was shown ready assembled. The advertisement was seen by a customer who later went to see the unit in store, where it was on display also ready assembled. The customer agreed to purchase the unit, but when it was delivered he found that it was in sections and that he had to assemble it himself. The advertisement was held to be a false trade description in that it gave a false description of the composition of the goods. It is also possible under the Act for the description of goods to be accurate but still to give rise to an offence if that description is misleading. In *Dixons Ltd* v. *Barnett* (1988)[11] a telescope bore the clear statement that it magnified up to 455 times. This was true, but in fact the telescope had a maximum useful magnification of 120 times: beyond that, the image became less clear and became no clearer with higher magnification. The shop was convicted because, although the statement as to magnification was true, it was misleading.

Strict liability

An offence is committed under this part of the Act regardless of the absence of any blame on the part of the person making the false trade description. Its falsity is enough to secure the commission of an offence. This makes the offence one known as a 'strict liability' offence[12].

Due diligence defence

However, the Act does provide for what is called the 'due diligence' defence. This allows the defendant to escape conviction if he can show that he took all reasonable precautions and exercised all due diligence to avoid commission of the offence. The cases show that this is a very difficult defence to satisfy. In the case of *Hicks* v. *Sullam*[13] bulbs were falsely described as 'safe'. The bulbs had been imported from Taiwan. There were 110 000 in all. None had been sampled to test for their safety and no independent test reports were obtained. The defendant's agent in the Far East had checked the bulbs and had reported no defects. The court ruled that the defence had not been made out.

While most defendants failed to make out a defence, some do occasionally succeed. In one case[14], a supplier was charged under the Act with falsely describing jeans as being the manufacture of Levi Strauss. The jeans had been obtained by him from a business associate in Greece with whom he had dealt for a couple of years. They were sold to him for £1 to £2 less than normal wholesale price. He examined the goods and they appeared to him to be in order. It was held that the defence had been made out. Similarly, in *Tesco Supermarkets Ltd* v. *Nattrass*[15], the defence was made out when the defendants showed they had devised a proper in-store system for ensuring compliance with the Act, and that they had done all they reasonably could, despite its breakdown on the particular occasion, to ensure that the system was implemented by the staff.

6.1.1.1 Statements as to services, facilities and accommodation

Section 14 of the Trades Description Act makes separate provision in relation to false or misleading statements as to services, facilities or accommodation. It is an offence for any person in the course of trade or business to make a statement which he knows to be false, or recklessly to make a statement which is false, on any of the following:

- the provision of any services, accommodation or facilities;
- the nature of any services, accommodation or facilities so provided;
- the time at which, manner in which or persons by whom any services, accommodation or facilities are provided;
- the examination, approval or evaluation by any person of any services, accommodation or facilities so provided;
- the location of services, accommodation or facilities.

The statement of offence needs to assert only that the defendant is charged with recklessly making a false statement and in what way it is false. Cases[16] have shown that to specify which subparagraph of s. 14(1) is contravened is unnecessary and likely to result in complications. The interpretation of 'service' has been clarified in cases[17,18,19] and means doing something for someone; while a facility means providing the opportunity or wherewithal for someone to do something for himself. By guaranteeing a refund on the price of a book containing instructions for a gambling system, Ashley[20] made a statement as to the nature of the services. Similarly, in a timeshare presentation, Global Marketing Europe (UK) Ltd[21] also made a statement as to service to be provided.

The Act applies only to statements of fact and not to promises which cannot be said to be true or false when made: *Beckett v. Cohen*[22]; *R. v. Sunair Holidays Ltd*[23]. The Act, however, may extend to implied statements of present intention, means or belief: *British Airways Board v. Taylor*[24]. Statements about services provided in the past also fall within the Act: *R. v. Bevelectric*[25]. In *Roberts v. Leonard*[26] it was held that the provisions of s.1 of the Act applied to the professions and it can safely be assumed that this is now the case with s.14.

A statement is false if false to a material degree, and anything likely to be taken as a statement covering the matters referred to above would be false if deemed to be a false statement of the relevant matter [s.14(2)a and (4)].

The mental element

In contrast to the position with false trade descriptions, the offence created by s. 14(1) is conditional upon the party charged knowing that the statement was false or making it recklessly. A statement is made 'recklessly' if it is made 'regardless of whether it is true of false whether or not the person making it had reason for believing that it may be false' [s.14(2)b]. To prove recklessness, it is enough to show that the party charged did not have regard to the truth or falsity of a particular statement, even though it cannot be said that he deliberately closed his eyes to the truth or had any kind of dishonesty in mind: *MFI Warehouses Ltd v. Nattrass*[27].

Where a person has no knowledge of the falsity of a statement at the time of its publication in a brochure, but did when the statement was read by a complainant, the statement would be a false statement every time business was done on the basis of the incorrect brochure: *Wings Ltd* v. *Ellis*[28]. In theory, the due diligence defence applies equally to offences under s. 14, but, since this section requires knowledge or recklessness before an offence is committed, in practical terms the defence will not generally be applicable.

Disclaimers

The courts have also been prepared to allow the use of disclaimers to avoid conviction, but have insisted that the disclaimer will be effective only if it is as 'bold, precise and compelling' as the false description it is attempting to disclaim. This is laid down in *Norman* v. *Bennett*[29] where a car dealer sought to disclaim a false mileometer reading with the statement 'speedometer reading not guaranteed'. This was contained in the small print of the contract and was held to be ineffective. In contrast, it was held in *Newham London Borough* v. *Singh*[30] that a disclaimer was effective when it was placed over the mileometer and read 'Trade Descriptions Act 1968. Dealers are often unable to guarantee the mileage of a used car on sale. Please disregard the recorded mileage on this vehicle and accept this as an incorrect reading'.

In *R.* v. *Bull*[31], a statement in a sales invoice alongside an odometer reading stating that the reading had not been confirmed and must be considered incorrect was held to be a valid disclaimer, while in *R.* v. *Kent County Council*[32] a trader who sold counterfeit goods, but who posted disclaimers and advised customers that the goods were copies, was also held to have used a valid disclaimer.

Penalties

The penalties for breach of the Act depend on the court the case is brought in. Most cases are brought in magistrates' courts where the penalty is a maximum fine of £5000. More serious cases are brought in the Crown Court where the penalty is a fine of any amount, a maximum sentence of two years, or a combination of both. In addition, the Powers of Criminal Courts Act 1973[33] empowers the court to award compensation to consumers affected by the breach of the Trade Descriptions Act.

6.1.2 Pricing offences

The Consumer Protection Act 1987[4]. makes it an offence for a price to be indicated which is misleading as to the price at which any goods, services, accommodation or facilities are available. In *MFI Furniture Centres Ltd* v. *Hibbert*[34] the prosecution did not have to produce an individual consumer to whom a misleading price was given. It was held in *Toys R Us* v. *Gloucestershire County Council*[35] that an indication is not misleading when the goods carry a ticket price lower than that shown when the bar code is run through the till if the retailer has a policy of always charging the lower price.

The Code of Practice

The novel feature of the Act is that it provided for the adoption by the Secretary of State for Trade and Industry, after approval by Parliament, of a code of practice which gave practical guidance to traders on how price indications should be given to avoid the commission of an offence. A Code of Practice for Traders on Price Indications[36] has been adopted.

It is important to understand that the Code is not binding on traders. A contravention of the Code is expressly said by the Act not of itself to give rise to an offence, but can be used as evidence that an offence in fact has been committed. Similarly, a trader who applies the Code cannot be entirely certain in law that his price indication is not misleading, although such compliance will again be evidence that his pricing is indeed not misleading. In practice, however, it will be very unusual for the presumptions raised by compliance with, or breach of, the Code to be displaced.

In a case[37] on this point, Stanley Ltd advertised, on 2 April 1992, an occasional table for £7.99. On 14 October, it was reduced to £4.99 in a 'Style and Value' promotion. There was a point of sales notice on or near the tables stating 'Style and Value, Occasional Table, Chipboard, save £3, now £4.99 (in large numbers) was £7.99'. The tables continued to be advertised at this price until 12 March 1993. In the mean time, on 10 November 1992, the tables were advertised in the press as a 13 day event headed 'SUPER SAVERS 13 DAY EVENT MUST END TUES 24'. The advertisement stated in the top right-hand corner 'SAVE £3'. In the bottom of the right-hand side was a picture of the table and the body of the advertisement read 'OCCASIONAL TABLE WAS £7.99. NOW ONLY £4.99'.

Around 24 November, the 13 day event ended but, contrary to what the advertisement had said, the table continued to be priced at £4.99 and remained in the 'Style and Value' that it had previously been in before the 13 day event. On 27 November, a customer bought one of the tables at the advertised price of £4.99. On 21 December the Christmas sale began with the table still priced at £4.99 but with a slightly different point of sale notice headed 'SALE, ROUND OCCASIONAL TABLE, NOW £4.99 WAS £7.99'. On 30 December, the same purchaser returned and bought three more of the tables for £4.99. On 12 March 1993 the price of the table was increased to £7.99. Two informations were laid against the company alleging offences under the Act. The first arose from the newspaper advertisement of 10 November, the second from the point of sale notice. The court held that the comparisons made by the advertising were misleading and did not accept that the Code of Practice was ambiguous. The meaning of the Code was clear and, when construed in the context of the legislation, there was no problem in concluding what it meant.

Due diligence defence

A due diligence defence on the lines of that set out in relation to the Trade Descriptions Act applies.

Penalties

The penalty for misleading pricing is a maximum fine not exceeding £5000 where the case is prosecuted in a magistrates' court. More serious cases are taken to the Crown Court where the penalty is a fine of unlimited amount. There is no power to impose a custodial sentence for pricing offences. Under the Powers of Criminal Courts Act 1973 the courts may make compensation orders in favour of the victim of a pricing offence.

6.1.2.1 Price indications

As well as the general ban on misleading price claims discussed above, there are also enactments imposing positive duties as to price indications. The Price Indications (Method of Payment) Regulations 1990[38] apply to traders who give indications of price for goods, services, accommodation or facilities and who charge different prices for different methods of payment. The Regulations do not apply to motor fuel.

The Price Marking Order 1991[39] applies to goods sold by retail, and to advertisements for such goods, whether the goods are sold in shops, by mail order or by other methods. The Order prescribes unit pricing for goods sold from bulk or pre-packed in variable quantities and for goods sold in pre-established quantity packs. Prices should be the final price, inclusive of VAT though this is relaxed where sales are mainly to businesses and, where goods or services have to paid for at the same time, this should be spelled out. Special provision is made for the indication of the price of motor fuel, jewellery and precious metals, and for indicating general price reductions.

In *Allen* v. *Redbridge*[40] the court held that the Order does not require a retailer to place a price label in front of an article or to put the article in a position where it could be handled by a prospective buyer. It was not an offence when goods bearing price labels on the bottom or the back of boxes containing the goods were stored in a locked cupboard to which access was available only with the help of a shop assistant.

Finally, there are the Price Indications (Bureaux de Change) (No. 2) Regulations 1992[41]. These apply to any trader carrying on the business of selling foreign currency in exchange for sterling. Clear and accurate information must be given on the buying and selling rates, terms of business, commissions and other charges. Receipts must be given setting out the terms unless the transaction is made by machine.

6.1.3 Truth in lending

The Consumer Credit Act 1974[42] requires that credit and hire advertising is not misleading. A building society[43] offered 'low start' mortgages. While the normal period of a loan was 25 years, under the 'low start' arrangement a borrower would pay 1% interest for 6 months, 2% under the society's prevailing rate for the next 6 months and 0.5% less than the society's rate for the next year. Thereafter the rate would be the society's current rate. At the time of the advertisement, the current rate was 8.45%. The advertisement showed an annual percentage rate (APR) of charge of 1.1% variable. The prosecution claimed that the APR had not been calculated in accordance with the Consumer Credit (Total Charge for Credit) Regulations 1980[44]. These require that a calculation which provides for a variation dependent upon the occurrence of an event shall be based on the assumption that the particular event will not occur. The Regulations exclude from the definition of 'event' something which is certain to occur. The society argued that the Regulations required the APR to be calculated on the assumption that the initial interest rate of 1% would not change during the rest of the term of the mortgage unless the rate was 'certain' to change at the end of the low rate period. The court ruled that the assumption that the interest rate would not change in circumstances where it was certain to change was misleading. The test to apply was to ask whether there was a real or realistic possibility that at the end of the initial period the rate would be exactly 1%, or whether the chance of that was so remote that it should be disregarded. It held that this was so remote a chance that it could be ignored and the advertisement was, therefore, misleading.

The somewhat complicated provisions of the Consumer Credit (Advertisements) Regulations 1989[45] provide that a credit and hire advertisement must be in one of three categories, namely simple, intermediate or full. A simple advertisement will do little more than indicate the name of the advertiser and the nature of his business, while a full advertisement will often give an example of repayment terms and the APR.

6.2 A fair quality of goods and services

The Sale of Goods Act 1979[46] imposes a number of obligations on sellers of goods. First, he must provide goods which correspond to whatever description has been given; second, the goods must be reasonably (not absolutely) fit for their purpose, and lastly, the seller must provide the buyer with goods which are of 'satisfactory quality'. Prior to the amendment of the Act in 1994[47], the requirement was that the goods be of 'merchantable quality'. Goods are of satisfactory quality 'if they meet the standard that a reasonable person would regard as satisfactory, taking account of any description of the goods, the price (if relevant) and all other relevant circumstances'. 'Quality' refers to the state and condition of the goods and includes the following aspects:

- fitness for all the purposes for which the goods are commonly supplied
- appearance and finish
- freedom from minor defects
- safety and durability.

In the event of a breach of any of the foregoing provisions, the buyer will be entitled to reject the goods and to claim damages for any loss suffered.

He will also be entitled to the return of the purchase price. If the goods had been bought by the use of a credit card and cost more than £100 but not more than £30 000, the Consumer Credit Act gives him the choice of bringing his action against the credit card company as an alternative to suing the seller.

The Sale of Goods Act only covers contracts of sale. Comparable duties are imposed in contracts which are closely related to sale under the provisions of the Supply of Goods (Implied Terms) Act 1973[48] (hire-purchase contracts) and the Supply of Goods and Services Act 1982[49] (contracts of hire or contracts where services and goods are supplied, such as the installation by a gas company of central heating).

Services

The Supply of Goods and Services Act requires that all who offer a service must provide that service with reasonable care and skill. It also states that if no time is agreed for the performance of the service, it must be carried out within a reasonable time, and if no charge for the service is agreed in advance a reasonable price must be paid. The Act expressly allows anyone providing a service to assume a stricter duty in respect of skill, care and time of performance.

6.3 Product safe...

The general safety ... duct
Safety Regulations ... er or
distributor to mark ... ct' or
which is a 'dangero ... s not
'safe'. In turn a 'sal ... eable
conditions of use i ... only a
minimum risk cor ...

The general saf ...

- second-hand p ...
- those supplied ... ...r .. .cconditioning before use provided that the buyer is so informed
- any product whose safety is already covered by specific EU legislation.

Producers are required to provide information to enable consumers to assess risks inherent in the product throughout its expected life and to take appropriate precautions. The producer should mark the product to identify batches, carry out sample testing, investigate complaints, keep records of feedback from retailers and have in place a product recall mechanism. Breach of the General Product Safety Regulations is a summary offence attracting a fine not exceeding £5000, a custodial sentence up to 3 months or a combination.

Consumer Protection Act 1987

Under s. 10(1) of this Act an offence is committed in relation to the supply, offer or agreement to supply, or the exposure or possession for the purpose of supply, of any goods when those goods fail to meet the general safety requirement. Section 10(2) defines the safety requirements as meaning that the goods 'are reasonably safe having regard to all the circumstances'. These circumstances include:

- the marketing, get up, use of any mark in relation to the goods and any instruction or warnings as to the keeping, use or consumption of the goods
- any published safety standards
- the existence of any means by which it would have been reasonable for the goods to be made safer.

In turn 'safe' is defined in s. 19(1) as meaning:

- that there is no risk or no risk apart from the minimum
- that none of the following will cause death or personal injury:
  - the goods
  - the keeping, use or consumption of the goods
  - the assembly of the goods
  - any emission or leakage from the goods or, as a result of the keeping, use or consumption of the goods, from anything else
  - reliance on the accuracy of any measurement, calculation or other reading made of the goods.

A person guilty of a breach of these general safety requirements is liable, on sentence, to a fine not exceeding £5000, a custodial sentence up to 6 months or a combination.

Since the General Product Safety Regulations came into effect, and because they include the disapplication of the general safety provisions in the Act in relation to products placed on the market by producers and distributors, those requirements of the 1987 Act have had diminishing importance. As a result, virtually all cases involving a breach of the general safety requirements are taken under the Regulations. However, the Act continues to apply where the supplier of the goods is not the person who first places the goods on the market.

6.3.1 Information exchange

Under the General Product Safety Directive if a Member State adopts emergency measures to prevent, restrict or impose specific conditions on the marketing or use, within its territory, of a product or product batch because of a serious and immediate risk to the health and safety of consumers, it must immediately inform the Commission, unless the effects of the risk do not or cannot extend beyond the territory of that Member State. The Commission will immediately inform the other

Member States who will inform the Commission of any measures adopted to counter the risk. The Commission can intervene directly if it becomes aware through notification or information from the Member States of the existence of a serious and immediate risk from a product, and if:

- one or more Member States have adopted restrictions on its marketing, or withdrawn it from the market
- Member States do not agree on the measures to be taken
- The risk cannot be dealt with adequately under other specific EU legislation relating to the product, and
- the risk can only be eliminated by adopting Community measures.

The EU Council can then adopt a Decision, after consulting the Member States, requiring them to take certain temporary measures.

The Commission is supported in this area by a Committee on Product Safety Emergencies, which comprises representatives from the Member States and is chaired by the Commission. This committee is consulted on draft measures to be taken and must give its opinion, by weighted majority, within a time limit set by the chairman according to the urgency of the situation, but in any case in less than 1 month. The Commission will accept the measures recommended by the committee and submit them to Council for adoption. If the Council has not acted within 15 days, the Commission will act on the recommendations. Measures adopted under this procedure are only valid for 3 months but the period can be extended. Member States must implement the agreed measures within 10 days. The competent authorities of the Member States must allow the parties concerned an opportunity to give their views and will inform the Commission accordingly. In the UK, this system operates through the Consumer Safety Unit of the DTI. Pharmaceuticals, animals, products of animal origin and radiological emergencies are excluded from the rapid notification system since they are dealt with under separate EU legislation.

6.3.2 Product recall

Article 6(h) of the General Product Safety Directive requires EU Member States to have in place appropriate measures to ensure the effective and immediate withdrawal of a dangerous product and, if necessary, that product's destruction. No specific UK legislation has been introduced to implement this requirement, but it is dealt with by provisions of the Consumer Protection Act through prohibition notices, notices to warn, suspension and forfeiture orders (see section 6.3.3 below).

In the motor industry, the recall of motor vehicles is dealt with by a voluntary set of arrangements agreed between the Department of Transport and the industry which operates through the DVLA. Failure of a manufacturer to recall dangerous products can lead to an action against him in negligence: *Walton v. British Leyland UK Ltd*[50].

6.3.3 Notices and orders

The Consumer Protection Act 1987 empowered the enforcing authorities to issue various notices and orders.

6.3.3.1 Prohibition notices

These notices are addressed to individual traders requiring them to cease the supply of the specified goods which the Secretary of State considers to be unsafe. For example, a prohibition notice was issued to a shopkeeper requiring him to stop supplying certain rice cookers which were found to be unsafe. Notices have also been issued in respect of elastic sweet-like toys on the ground that they presented a potential choking hazard to children. Another case involved the withdrawal from the market of mercury soap as a skin lightener.

6.3.3.2 Notices to warn

A notice to warn may be issued by the Secretary of State and require the person on whom it is served to publish, at his own expense, warnings in the form, manner and at times specified in the notice, that he supplies or has supplied unsafe goods. To date no such notices have been issued but it is not unusual for manufacturers voluntarily to advertise in the press that products identified in the advertisement are unsafe and to request their return for repair or refund.

6.3.3.3 Suspension notices

Local trading standards officers are empowered to issue suspension notices where they have reasonable grounds for supposing that there is a breach of the general safety requirement. The maximum duration of the notice is 6 months and during that period the person on whom it is served cannot deal in the goods without the consent of the local trading standards authority

A person guilty of a breach of any of the above notices can be liable to a fine of up to £5000, a custodial sentence up to 6 months or a combination.

6.3.3.4 Forfeiture orders

Local trading standard officers may apply, under s.15(2)b, for the forfeiture of goods on the grounds of a breach of either a general safety requirement, a prohition notice, a notice to warn or a suspension notice. Instead of ordering the destruction of the goods, the court may direct that the goods be released to a person nominated by the court provided that person only supplies them either to a trader whose business is that of repairing or reconditioning such goods or to one who receives them as scrap.

6.3.4 Safety regulations

Power is given by s.1(1) of the 1987 Act to the Secretary of State to make safety regulations whose purpose must be to ensure that goods are safe or that goods which are, or in the hands of a person of a particular description are, unsafe are not made available generally or to persons of that description. Also that appropriate information is, or inappropriate information is not, provided in relation to the goods. There is a considerable body of such regulations, many were made under previous legislation but are now treated as if made under the 1987 Act. Information about these regulations can be obtained from the Consumer Safety Unit, Department of Trade and Industry, Victoria Street, London SW1H 0ET.

6.3.5 Food and medicines

The Food Safety Act 1990[51] makes it an offence to sell food not conforming to the 'food safety requirements'. This is more narrowly defined in the general safety requirement discussed above. Food is deemed to breach the food safety requirement if it is unfit for human consumption, if it has been rendered injurious to health by any of the operations specified in the Food Safety Act, or if it is so contaminated that it would not be reasonable to expect it to be used for human consumption. Food is not covered by the general safety requirement laid down in the Consumer Protection Act, but it is within the general safety requirement contained in the General Product Safety Regulations.

The Medicine Act 1968[52] imposes strict controls on the manufacture and supply of medicinal products. In particular, most such products will require a 'product licence', while some will be available on prescription only.

6.4 Product liability

Part I of the Consumer Protection Act created what is called a system of 'strict liability' for defective products, allowing an injured person to sue without the need to prove negligence. It had long been regarded as anomalous that the person who bought defective goods could sue under the Sale of Goods Act for any injury caused without the need to prove negligence, whereas a non-purchaser (e.g. a bystander, a member of the purchaser's family or a person to whom the goods had been given as a gift) could only recover damages if he could prove negligence.

Part I of the Act now provides that damages can be recovered simply on proof that the product is defective, which means that its safety is not such as persons generally are entitled to expect. Liability is placed on the producer, which in this context is stated to include any person who 'own-brands' a product as though he were in fact the producer; and the first importer of the product into the EC. The actual supplier of the defective product can also be made liable, but this can only be where he has been asked by the injured party to name the actual producer and he fails to comply with the request or identify the party who supplied him with the goods within a reasonable time.

The producer of a defective product will not be liable in every case. For instance, he will not be liable if he can show that the goods were not defective when supplied by him. Again, the producer of a defective component has a defence if he can show that he was following the instructions of the producer of the product which was to incorporate the component; or if the defect was due to the design of the end-product.

The most contentious defence contained in the Act is usually called the 'development risks' defence. Under this defence, the producer of a defective product will have a defence if he can show that the 'state of scientific and technical knowledge at the relevant time was not such that a producer of products of the same description might be expected to have discovered the defect if it had existed in his products while they were under his control'. The Act implemented the provisions of an EU directive under which the continuing existence of this defence was to be reviewed. That review has now taken place and has recommended the removal of this defence but no formal decision has been adopted.

The Act applies to damage to property in exactly the same way as it applies to personal injury. However, actions for damage to property cannot be brought unless the claim exceeds £275. If it does exceed that amount, then the whole of the loss can be claimed.

All claims for damage, whether to person or property, must be brought within three years, but no claims can be brought at all once 10 years has expired from the time of supply of the defective product.

6.5 Misleading advertising

In enacting the Control of Misleading Advertisements Regulations 1988[3], the UK adopted Council Directive 84/450. Under the Regulations, the Director General of Fair Trading may seek a court injunction against an advertisement where a complaint has been brought that it is misleading. In deciding whether to apply for a court injunction, he must first consider whether the advertisement in question has been the subject of complaint to 'established means' of dealing with such complaints. This is a reference to such bodies as the Advertising Standards Authority whose British Code of Practice[53] represents the industry's efforts at self-regulation. The Office of Fair Trading (OFT) obtained an injunction in the case of *Director General of Fair Trading* v. *Tyler Barrett and Co. Ltd*[54] in which the court was presented with evidence that advertisements by this company were misleading and false in a number of respects. The injunction prevented the continuation of advertisements, in the form of telephone cold-calling and personal visits by representatives of the company to the premises of small local businesses offering to obtain business grants from the EU for a search fee of £350 plus 10% of any grant obtained. Clients were led to believe that the company would assist them in obtaining grant funding for a variety of business purposes. The availability of such grants and the likelihood of success in obtaining such funding were greatly exaggerated by the company. Clients were reassured that, should no grant be obtained for them, they would receive a full refund of their search fee.

The information given to the clients regarding the availability of grants was incorrect and what clients actually received was a standardised list of grant making bodies, most of which was wholly irrelevant to the client's needs. When clients realised the very limited nature of the service and that they had been deceived, they sought to obtain the refund without success. The information provided by the company to small businesses and its failure to provide a refund of the search fee have been the subject of many complaints to trading standards departments and other bodies, such as local business link offices. To comply with the injunction, Tyler Barrett and Co. Ltd and its director Peter Kemp will have to stop making misleading statements about what they offer to their clients. If they wish to get the injunction lifted, they will have to attend a further hearing.

Instead of taking court action, the OFT can obtain assurances from the advertiser that a particular course of advertising will stop. One advertiser sent unsolicited mailshots which used words like: 'Would you like more money in your pocket?', 'Too busy earning a living to make money?', 'Looking for a genuine chance to improve your lifestyle?'. Attached to them were handwritten self-adhesive notes bearing the comment 'Working wonders for me ... knew you'd be interested' followed by the initial 'M'. The OFT considered the advertising was misleading in that it suggested personal recommendations. The advertiser agreed to give the Director General an undertaking[55] that he would not put out advertisements which gave consumers the false impression that they were sent by, or a product was recommended by, somebody they knew.

Broadcast advertising

Under the Regulations, the Director General has no power in relation to advertisements carried on any television, radio or cable service. Where an advertisement is misleading, neither the Independent Television Commission nor the Radio Authority have power to seek an injunction, instead they can refuse to transmit the advertisment.

6.6 Exclusion clauses

At one time, particularly in the field of the sale of goods, the small print of the contract would often contain clauses, usually called 'exclusion clauses', which took away from the consumer the rights given him under such legislation as the Sale of Goods Act. The use of such clauses is now subject to the controls imposed by the Unfair Contract Terms Act 1977[56] and the Unfair Terms in Consumer Contracts Regulations 1994[6].

6.6.1 The Unfair Contract Terms Act 1977

In sales to a consumer, the Act does not allow the seller to avoid the obligations which are imposed on him by the Sale of Goods Act (see section 6.2 above). Even to try to avoid the obligation is a criminal offence under the provisions of the Consumer Transactions (Restrictions on Statements) Order 1976[57].

In the case of sales to other businesses, exclusion clauses will be effective, but only if they are reasonable. Similar constraints are imposed in relation to contracts where possession or ownership of goods passes, but the contract is not one of sale or hire-purchase.

The same Act also controls the operation of other types of exclusion clause when incorporated into a consumer contract or in a contract which is on written standard terms. The principle is that three types of clause which might be used in such a contract are valid only if they can be proved to be reasonable; such clauses are those which:

- seek to allow a person not to perform the contract;
- seek to allow him to provide a performance 'substantially different' from that which was reasonably expected; and
- allow the person in breach of contract to be free of all liability to pay damages for his breach.

Suppose that a term in a holiday contract says that a person may have to share a room if the tour operator so decides instead of getting a single room that he has booked. This will be a clause seeking to provide a performance substantially different from that reasonably expected. Under the Act, this clause will not be valid unless the tour operator can prove it was reasonable. Where a clause is valid only if reasonable, the assumption laid down in the Act is that the clause is unreasonable until the contrary is proved.

6.6.2 The Unfair Terms In Consumer Contracts Regulations 1994

Unlike the Unfair Contract Terms Act, the Regulations do not automatically render any terms void, but the range of clauses dealt with is wider than in the Act. At the same time the scope of the Regulations is narrower because they only cover consumer contracts. Apart from certain excluded contracts, the Regulations apply to all contracts made between an individual and a business which have not been individually negotiated.

The contracts excluded by schedule 1 are contracts of employment, succession rights, family law, company law and any term required by UK law or by international convention. A term is not regarded as 'individually negotiated' if it was drafted in advance without the individual's involvement. Notwithstanding that a specific term, or certain aspects of it, in a contract has been the subject of individual negotiation, the Regulations will apply to the rest of the contract if an overall assessment of the contract indicates that it was a pre-formulated contract.

Contracts must be in 'plain intelligible language'. If not, and there is doubt about a written term, the interpretation most favourable to the individual will prevail. An unfair term is not binding on the individual although the contract will remain binding if it can operate without the unfair term.

A term is unfair if, contrary to the requirements of good faith, it causes a significant imbalance of rights and obligations to the detriment of the consumer. Schedule 2 states that, in assessing good faith regard shall be had to the strength of the bargaining position of the parties; whether the individual had an inducement to agree to a particular term; whether the goods were sold or supplied to the special order of the individual, and the extent to which the seller or supplier has dealt fairly and equitably with the individual. Schedule 3 includes what it calls an indicative and illustrative list of terms which may be regarded as unfair. These include terms requiring a consumer who fails to fulfil his obligations, to pay a disproportionately high sum in compensation or which irrevocably binds a consumer to terms with which he had no real opportunity of becoming acquainted before the making of the contract. The Regulations go on to state that, provided it is in plain intelligible language, no assessment shall be made of the fairness of any term which defines the subject matter of the contract or which concerns the adequacy of the price or remuneration to be paid for the goods or services involved.

Complaints about unfair terms are dealt with by the OFT which can obtain undertakings and or seek injunctions in respect of the unfair terms. One clause that was considered unfair was contained in a contract used by Yorkshire Gas Show Rooms Ltd[58]. This guaranteed installations so long as the invoice price was paid in full on completion of the work. The effect of this was to prevent the consumer from exercising a right of retaining part of the payment in respect of any claim they might have against the company because, if they did so, they would forfeit their right to the guarantee. If payment was just one day late, the guarantee would be invalidated. This was considered to be unfair and the term was amended to provide that the guarantee would come into effect only after payment had been made in full. This gave time for the settlement of disputes and for defects in the installation to be remedied.

The OFT has established a special unit to deal with unfair terms and publishes regular bulletins. Anyone wishing to make a complaint should write to the Unfair Contract Terms Unit, Room 505, Field House, Bream's Buildings, London EC4A 1PR.

6.7 Consumer redress

Up until 1973 going to court in pursuance of a consumer claim could be a daunting and expensive business. It was then that a small claims or arbitration procedure was set up which operated through the County Court. Any claim within the County Court jurisdiction (generally up to £50 000) can be referred on application to an arbitration heard by the District or County Court judge or even by an outside arbitrator. Any such arbitration has the effect of a full County Court judgment, though it is usually heard in private by the arbitrator in an informal manner and without the normal rules of court procedure applying. If the sum claimed does not exceed £3000, it will go to arbitration automatically if either party desires. Above that limit, both parties will have to agree before the matter can use this procedure. An important feature of the arbitration system is that the rule as to costs has been considerably modified. The loser of a case normally will be asked to pay only a nominal sum. As a rule, he will not have to pay anything in respect of his opponent's legal fees.

References

1. Final report of the Committee on Consumer Protection, Cmnd 1781, HMSO, London (1962)
2. *Trade Descriptions Act 1968*, HMSO, London (1968)
3. *The Control of Misleading Advertisements Regulations 1988*, SI 1988 No. 915, HMSO, London (1988)
4. *Consumer Protection Act 1987*, HMSO, London (1987)
5. *The General Product Safety Regulations 1994*, SI 1994 No. 2328, HMSO, London (1994)
6. *The Unfair Terms in Consumer Contracts Regulations 1994*, SI 1994 No. 3159, HMSO, London (1994)
7. *Misrepresentation Act 1967*, HMSO, London (1967)
8. Cadbury Ltd *v.* Halliday [1975] 2 All ER 226
9. British Gas Board *v.* Lubbock [1974] 1 WLR 37
10. Queensway Discount Warehouses *v.* Burke [1985] BTLC 43
11. Dixons Ltd *v.* Barnett [1988] BTLC 311
12. Alec Norman Garages Ltd *v.* Phillips [1984] JP 741
13. Hicks *v.* Sullam [1983] MR 122
14. Westminster City Council *v.* Pierglow Ltd (8 February 1994, unreported)
15. Tesco Supermarkets *v.* Nattrass [1972] AC 153
16. Regina *v.* Piper [1995] 160 JP 116
17. Newell *v.* Hicks [1983] 148 JP 308
18. Kinchin *v.* Ashton Park Scooters [1984] 148 JP 540
19. Dixons Ltd *v.* Roberts [1984] 148 JP 513
20. Ashley *v.* Sutton London Borough Council [1994] 159 JP 631
21. Global Marketing Europe (UK) Ltd *v.* Berkshire County Council Department of Trading Standards [1995] Crim LR 431
22. Beckett *v.* Cohen [1973] 1 All ER 120
23. Regina *v.* Sunair Holidays Ltd [1973] 2 All ER 1233
24. British Airways Board *v.* Taylor [1976] 1 All ER 65
25. Regina *v.* Bevelectric [1992] 157 JP 323
26. Roberts *v.* Leonard [1995] 159 JP 711
27. MFI Warehouses Ltd *v.* Nattrass [1973] 1 All ER 762
28. Wings Ltd *v.* Ellis [1985] AC 272
29. Norman *v.* Bennett [1974] 3 All ER 351
30. Newham London Borough *v.* Singh [1987] 152 JP 239
31. Regina *v.* Bull, *The Times*, 4 December 1993
32. Regina *v.* Kent County Council (6 May 1993, unreported)
33. *Powers of the Criminal Courts Act 1973*, HMSO, London (1973)
34. MFI Furniture Centres Ltd *v.* Hibbert [1995] 160 JP 178
35. Toys R Us *v.* Gloucestershire County Council [1994] 158 JP 338
36. *Code of Practice for Traders on Price Indications*, HMSO, London
37. AG Stanley Ltd *v.* Surrey County Council [1994] 159 JP 691
38. *The Price Indications (Method of Payment) Regulations 1990*, SI 1990 No. 199, HMSO, London (1990)
39. *The Price Marking Order 1991*, SI 1991 No. 1382, HMSO, London (1991)
40. Allen *v.* Redbridge [1994] 1 All ER 728

41. *The Price Indications (Bureaux de Change) (No. 2) Regulations 1992*, SI 1992 No. 737, HMSO, London (1992)
42. *The Consumer Credit Act 1974*, HMSO, London (1974)
43. Scarborough Building Society *v.* Humberside Trading Standards Department [1997] CCLR 47
44. *The Consumer Credit (Total Charge for Credit) Regulations 1980*, SI 1980 No. 51, HMSO, London (1980)
45. *The Consumer Credit (Advertisements) Regulations 1989*, SI 1989 No. 1125, HMSO, London (1989)
46. *The Sale of Goods Act 1979*, HMSO, London (1979)
47. *The Sale and Supply of Goods Act 1994*, HMSO, London (1994)
48. *The Supply of Goods (Implied Terms) Act 1973*, HMSO, London (1973)
49. *The Supply of Goods and Services Act 1982*, HMSO, London (1982)
50. Walton *v.* British Leyland (UK) Ltd (12 July 1978, unreported)
51. *The Food Safety Act 1990*, HMSO, London (1990)
52. *The Medicines Act 1968*, HMSO, London (1968)
53. Advertising Standards Authority, *British Code of Advertising Practice* (9th edn), Advertising Standards Authority, London (1995)
54. Director of Fair Trading *v.* Tyler Barrett and Co. Ltd (1 July 1997, unreported)
55. *Consumer Law Today*, December 1997
56. *The Unfair Contract Terms Act 1977*, HMSO, London (1977)
57. *The Consumer Transactions (Restrictions on Statements) Order 1976*, SI 1976 No. 1813, HMSO, London (1976)
58. Office of Fair Trading, 4th Bulletin on Unfair Contract Terms, December 1997, Office of Fair Trading, London (1997)

Further reading

Lawson, R.G. *Exclusion clauses and unfair contract terms*, FT Law and Tax, London (1995)
Abbott, H. *Product safety*, Sweet and Maxwell, London (1996)
Wright, C. *Product liability*, Blackstone Press Ltd, London (1989)
Consumer Law Today, published monthly by Monitor Press

Chapter 7
Insurance cover and compensation

A. West

7.1 Workmen's compensation and the State insurance scheme

The first Workmen's Compensation Act was passed in 1897 (eventually consolidated in the Workmen's Compensation Act 1925) and, as an alternative to a workman's rights at common law, imposed on the employer an obligation to pay compensation automatically in the event of a workman sustaining an accident in the course of his employment. There was no requirement of fault, the legislation being introduced to provide compensation where the workman was injured in purely accidental circumstances with no blame attaching to anyone and resembled therefore an insurance scheme. The system was operated with recourse to the County Court in the event of any dispute arising and facilitated a cheap and relatively quick payment of compensation. The amount of compensation was expressed as a weekly sum and was based on the average wage earnings during the previous 12 months with the employer whereas at common law if successful in establishing liability a workman was awarded a lump sum by way of damages. The workman did, however, have to elect between claiming at common law or claiming under the Workmen's Compensation Act.

Following the decision in *Young* v. *Bristol Aeroplane Company Limited* [1944] 2 All ER 293 it became established that a workman was precluded from pursuing a claim at common law even where he did not know of his right to elect if he had in fact accepted weekly payments under the Workmen's Compensation scheme. The Workmen's Compensation insurance policies issued at that time indemnified the insured against his liability to pay compensation under the Workmen's Compensation Act, the Employer's Liability Act 1880 and the Factories Act 1846 or at common law in the event of personal injury to any employee arising out of and in the course of his employment.

The introduction of the State scheme by the National Insurance (Industrial Injuries) Act 1946 can be considered as a compromise between the complete abolition of the common law system with its requirement of proof of fault on the part of the employer and the differing opinions of the type of accident insurance which would be most desirable.

Various types of benefits are available under the State insurance schemes for industrial injuries and are payable in respect of any person who has suffered personal injury caused by an accident arising out of and in the course of his employment or where such person suffers from what is termed a prescribed disease with reference to certain industrial occupations which may give rise to that particular disease. The phrases 'accident' and 'arising out of and in the course of his employment' have given rise to much dispute over the years since their introduction. An accident has been defined as an 'unlooked for mishap or untoward event which is not expected or designed' and by definition may be distinguished from a process involving, for example, repetitive movements of the hand or wrist which may give rise to a disability such as tenosynovitis where it is difficult to identify any particular event causing injury as opposed to considering the series of events as a whole forming a process.

There are many cases involving the question whether an act of an employee arises out of and in the course of his employment especially under the State insurance scheme and while these are beyond the scope of this text they may be studied in detail elsewhere[1]. For a comparatively recent decision on the topic illustrating some of the problem areas see *Nancollas* v. *Insurance Officer* and *Ball* v. *Insurance Officer* [1985] 1 All ER 833.

An employee suffering from the effects of an accident at work or from a prescribed disease may be entitled to a range of benefits determined by the current Social Security Act and supporting Regulations. The benefits may include:

1 *Statutory Sick Pay* – The Social Security and Housing Benefit Act 1982 introduced the concept of statutory sick pay (SSP) payable by the employer for the first eight weeks of absence due to injury or sickness. From 6 April 1986 the period of payment has been extended to 28 weeks. As from 6 April 1995 payment is made at one level subject to taxation. Previously it was paid at two rates based on average weekly earnings. Whilst receiving SSP there is no right to claim incapacity benefit.

2 *Industrial Injuries Disablement Benefit* – Where an employee becomes disabled as a result of an injury at work or as a result of one of the prescribed diseases, then he should qualify for Industrial Injuries Disablement Benefit. The requirements for payment of benefit are broadly loss of physical or mental capacity as a result of an industrial accident or disease. This means some impairment of the power to enjoy a normal life and includes disfigurement even though this causes no bodily handicap. The impairment assessment is expressed as a percentage subject to a maximum of 100% and originally no benefit was paid where the loss of faculty was less than 1%. However, since the introduction of Social Security (Industrial Injuries and Diseases) Miscellaneous Provisions Regulations 1986 entitlement to benefit only arises where the degree of disablement arising from the loss of faculty is assessed at 14% or more. But it is possible to receive Disablement Benefit if suffering from one of the prescribed diseases and disability is

1% or more, where aggregated assessments exceed 14%. Anyone suffering from disablement assessment between 14% and 19% will be paid at the 20% rate, payment taking the form of a weekly pension.
3. *Attendance Allowance/Constant Attendance Allowance* – Attendance allowance is paid under the National Insurance Scheme. Constant attendance allowance is paid under the Industrial Injuries Scheme and it is not possible to get both.
4. *Benefits for Prescribed Industrial Diseases/Benefits for Certain Specific Diseases* – The onset of certain industrial diseases or medical conditions is gradual or progressive and they fall outside the usual industrial injury benefit rules. A scheme of benefits was therefore created for 'prescribed diseases' (with qualifying conditions). See the Social Security (Industrial Injuries) (Prescribed Diseases) Regulations 1985.

The qualifying criteria for Constant Attendance Allowance is a serious handicap sufficient to require constant care and attention as a result of the effects of an industrial accident or disease.

To qualify for Attendance Allowance a person must be so severely disabled that they require frequent attention throughout the day in connection with bodily functions or continual supervision to avoid danger to themselves or others.

The qualifying feature for the payment of Mobility Allowance is the inability to walk a reasonable distance without distress or risk of damage to health.

The employee is not of course precluded from claiming other benefits where the absence from work does not arise from an industrial accident or prescribed disease. All that is required to be shown is that the claimant is incapable of work and he is so incapable if having regard to his age, education, experience, state of health and other personal factors there is no type of work which he can reasonably be expected to do. These benefits may include:

(a) *Incapacity Benefit* Incapacity benefit replaced sickness benefit and invalidity benefit in 1995 and accordingly has two elements. The short-term element pays a lower tax-free rate for the first 28 weeks of sickness to the self-employed or to those who do not qualify for SSP, a higher taxable rate is paid after 28 weeks, equal to the SSP rate. The long-term element is paid after 52 weeks, the basic rate being equal to the basic state retirement pension.
(b) *Severe Disablement Allowance* Claimants who do not qualify for incapacity benefit (because of insufficient National Insurance contributions) may be entitled to this allowance if they have been unable to work for 28 weeks provided they are assessed at 80% disabled unless aged under 20 when no assessment is necessary. Various disabilities can be aggregated for assessment[2].

Section 22 of the Social Security Act 1989 made provision for the Department of Social Security to collect from those paying compensation for injury or illness, the amount of benefit paid to persons as a result of such injury or illness. Effectively this entitled the Government to repayment of any State Insurance Scheme payments made to those injured or ill where those persons are entitled to compensation following pursuit of a common law claim. This Act has now been revised and is superseded by the Social Security (Recovery of Benefits) Act 1997.

7.2 Employer's liability insurance

Since 1 January 1972 it has been compulsory for employers to insure against their liability to pay damages for bodily injury or disease sustained by their employees arising out of and in the course of their employment. This was enacted by s.1(1) of the Employer's Liability (Compulsory Insurance) Act 1969 and failure to comply with the provisions of the Act by an employer renders him guilty of an offence and liable to summary conviction – s.5.

The Act contains a definition of the term 'employee' as including an individual who has entered into or works under a contract of service or apprenticeship with an employer whether by way of manual labour, clerical work or otherwise, whether such contract is expressed or implied or in writing – s. 2(1). Certain relatives of the employer are outside the ambit of the Act – s. 2(2)(a) – as are employees 'not ordinarily resident in Great Britain' – s. 2(2)(b).

The contract of insurance incorporates conditions compliance with which is itself a condition precedent to liability under the policy. Accordingly whilst an employer may incur liability to one of his employees, in the event of his failing to comply with a condition of the policy, for example failure to notify the insurer in reasonable time of an occurrence which may give rise to liability under the policy, the insurer may invoke non-compliance with the condition as a reason for refusing to indemnify the employer under the policy. In certain circumstances this could prejudice the injured employee's prospects of recovering damages. The Employer's Liability (Compulsory Insurance) General Regulations 1972 restrict the application of conditions in policies of insurance. The regulations do not, however, prejudice the rights of the insurer to recover from the policy holder sums which they have been required to pay by reason of application of the regulations. To ensure that employees are aware of the existence of the contract of insurance, ss. 5 and 6 of the Regulations deal with the requirement on the insurer to issue a certificate and the subsequent requirement on the employer for its display at his place of business in such a position as to be seen and read by every person employed whose claims may be the subject of indemnity under the policy. The Employer's Liability (Compulsory Insurance) Act 1969 is currently under review.

Policy Cover – the basic cover indemnifies the insured against liability at law for damages and claimant's costs and expenses in respect of bodily injury or disease sustained during the period of insurance by any person under a contract of service or apprenticeship with the insured whilst employed in or temporarily outside the territorial limits which are normally Great Britain, Northern Ireland, the Isle of Man or Channel Islands and arising out of and in the course of his employment. In view of the increased use of subcontract labour and to clarify the position regarding temporary staff and others working for an insured under various schemes and arrangements, the definition of employee has now been extended to include persons supplied to, hired or borrowed by the insured in the course of his business.

The criteria by which 'arising out of and in the course of his employment' is established are different in relation to Employer's Liability insurance and the State insurance scheme, the latter incorporating a broader definition. For an illustration of this aspect see *Vandyke* v. *Fender* [1970] 2 All ER 335 concerning the question of which insurer, motor or employer's liability, should deal with a claim where a company provides a car for its employees to go to or from work and an accident occurs on the road.

A more recent example of these issues is *Smith* v. *Stages* [1989] 1 All ER 833. Two employees were sent by their employers to carry out work at a site some distance from the site at which they had previously been working. They were paid 8 hours pay for the travelling time in addition to the equivalent of the rail fare, although no stipulation was made as to the mode of travel. On returning from the site the vehicle crashed killing the passenger. It was held that the employers were vicariously liable for the negligence of the driver. Both men were acting within the course of their employment when returning to their ordinary residence after completing the temporary work as they were travelling back in the employers' time and were paid wages and not merely a travelling allowance.

With effect from 31 December 1992 the Motor Vehicles (Compulsory Insurance) Regulations came into force requiring all passengers to be covered by motor insurance, including liability arising out of and in the course of employment.

The majority of all Employer's Liability claims emanate from accidents on the 'factory floor' often involving injuries sustained through contact with dangerous moving machinery. The Employer's Liability policy is designed to indemnify the employer against his legal liability to pay damages to employees for injuries sustained in such circumstances. This liability may arise either from the employers' breach of certain statutory duties or from a breach of their common law duties to their employees where the injured person can prove that the breach was causative of the injury. An example often encountered is the employer's duty to guard machinery referred to in ss. 12–14 of the Factories Act 1961. The duty to fence machinery securely is an absolute one and the fact that compliance with s. 14(1) of the Factories Act may render the machine unusable does not absolve the employers from their duty. This principle is illustrated by *Frost* v. *John Summers and Son Limited* [1955] 1 All ER 870 where a grinding wheel was held to have been a dangerous part of machinery within s.14(1). This decision was instrumental in bringing about special regulations (the Abrasive Wheels Regulations 1970) which enabled this type of machine to be operated without being in breach of the Factories Act 1961. Two other sections of the Factories Act, breaches of which are often alleged by injured workmen, are ss. 28 and 29 broadly encompassing the employer's duty to provide a safe place of work and safe means of access thereto; however, the duty is qualified by the words 'as far as is reasonably practicable'. The duty is not absolute, for example it may be impracticable to maintain passages in a condition such that there are never any slippery patches particularly after it has been raining – for example, see *Davies* v. *De Havilland Aircraft Company Limited* [1950] 2 All ER 582. This qualification on the duty of the employer effectively equates the statutory provisions containing the qualification with the employer's common law duty.

As a result of the Workplace Directive No. 89/654/EEC there is now greater emphasis on requiring employers to assess hazards and prevent injury. The directive sets out the framework for legislation required which has been implemented through the Workplace (Health, Safety and Welfare) Regulations 1992 (WHSWR) and associated Regulations, generally referred to as the 'six pack'.

The Management of Health and Safety at Work Regulations 1992 (MHSWR) are wide ranging and of a general nature and may overlap existing specific Regulations. Where duties do overlap, compliance with the specific Regulation will normally be sufficient to meet legislative requirements. However, where the duties under MHSWR go beyond the requirements of specific Regulations, additional measures will be needed to comply with MHSWR.

The 'six pack' of regulations replace many existing laws by repealing and revoking old legislation. For example:

1 Sections 12–16 of FA, relating to guarding of parts of machinery, are replaced by PUWER.
2 Sections 28 and 29 of FA, relating to the provision and maintenance of a safe workplace and access, are replaced by regs 5, 12, 13 and 17 of WHSWR.

At common law the employer's main duties are:

1 to take reasonable care that the place of work provided for the employee is safe, for example see *Quintas* v. *The National Smelting Co. Limited* [1961] 1 All ER 630,
2 to provide sufficient safe and suitable plant, for example see *Kilgullan* v. *W. Cooke and Co. Limited* [1956] 2 All ER 294,
3 to maintain such equipment, for example see *Henderson* v. *Henry E. Jenkins & Sons* [1969] 3 All ER 756, and
4 to provide a proper and safe system of work, for example see *General Cleaning Contractors Limited* v. *Christmas* [1952] 2 All ER 1110.

In addition an employer has an obligation to use care in the selection of fellow employees although this duty is less often encountered as a result of the development of the doctrine of vicarious liability whereby the employer will be liable for the negligent acts of his employees whilst acting in the course of their employment.

Any breach of these common law duties resulting in injury to an employee will give rise to liability against which the Employer's Liability policy may indemnify the insured in the event of damages being payable to the injured employee.

An insurer will on behalf of the employer, where applicable, raise a defence to a workman's claim. Various defences are available to him. These include the complete defences of:

1 *volenti non fit injuria* where the injured person has consented to run the risk,
2 'inevitable accident' where despite the exercise of reasonable care by the defendant the accident still occurred,
3 defences based on the Limitation Acts where the plaintiff fails to bring his action within the prescribed time limit, and
4 partial defences such as contributory negligence (see later text).

The defence of *volenti non fit injuria* has very limited application since the mere continuance in work that involves risk of injury does not imply acceptance of the risk of injury caused by the employer's negligence and this defence has rarely succeeded in circumstances of an injury to a servant by the negligence of his master. See, for example, *Bowater* v. *Rowley Regis Corporation* [1944] 1 All ER 465.

The onus of proving negligence or breach of statutory duty and that this failure was the cause of the accident rests on the plaintiff except where the facts of any accident are such that the accident would not have occurred without negligence. This is the doctrine of *res ipsa loquitur* whereby the defendant must prove that the accident could have occurred without negligence on his part, for example see *Scott* v. *London Dock Company* [1865] 3 H and C 596. For a more modern approach to this concept and a discussion of the problems involved see *Ward* v. *Tesco Stores* [1976] 1 All ER 219.

In the past, in contrast to the Public Liability policy, it was not usual to impose a limit of indemnity to the Employer's Liability policy. However, as from 1 January 1995 insurers have imposed a cap of £10 m per incident (£2 m for offshore risks). The Employer's Liability policy usually includes cover for all costs and expenses incurred with the insurance companies' consent and extends to include the cost of representation of the Insured at proceedings in a Court of Summary Jurisdiction arising out of an alleged breach of statutory duty resulting in bodily injury or disease which may be the subject of indemnity under the policy.

The phrase 'sustained during the period of insurance' is designed to pick up the disease risk even where the symptoms do not become manifest until many years later. Insurers are increasingly finding themselves facing claims relating to events which took place many years ago, a situation brought about because of the relaxation in the time limit for bringing claims, in particular the introduction of the 'disapplying' provisions inserted into the 1939 Limitation Act by the Limitation Act 1975 and consolidated by the Limitation Act 1980. These developments are highlighted in the case of *Buck* v. *English Electric Co. Limited* [1978] 1 All ER 271 where the widow of a man who died of pneumoconiosis was allowed to continue her husband's action for damages for personal injuries against his former employers despite the lapse of some 16 years between the deceased's knowledge of the onset of the disease and proceedings being commenced. An insurer, however, will only indemnify the insured for that part of the damages relating to the period for which the risk was held and during which there was causative exposure to the process to which the disease is in part attributable.

Claims for damages for noise-induced hearing loss are a prime example of retrospective liability giving rise to substantial difficulties for liability insurers. Deafness was added in 1975 to the list of prescribed industrial diseases under the Social Security (Industrial Injuries) (Prescribed Diseases) Regulations 1975. However, the right to benefit was limited to deafness caused by exposure to specific noise producing machinery within the metal manufacturing and shipbuilding industries, also requiring an exposure of 20 years or more within that industry. The qualifying occupations have been extended by subsequent regulations now consolidated within the Social Security (Industrial Injuries) (Prescribed Diseases) Regulations 1985.

The first reported case of an employee succeeding in a damages claim against his employer for deafness was *Berry* v. *Stone Manganese Marine Limited* [1972] 1 Lloyd's Reports 182 although the law has developed since that case and a more recent insight into the problems of a common law claim in this area can be gained from *McGuiness* v. *Kirkstall Forge Engineering Limited* QBD Liverpool 22 February 1979 (unreported). In this latter case the defendants were forgemasters and the plaintiff had worked for them for most of his working life operating a stamping press. The judge found that there was virtually no evidence that any employer in noisy industries was taking any steps at all to protect his workmen prior to the late 1950s and it was not until the late 1960s that anyone in the drop-forging industry began to show an interest in protecting workmen. The potential damage which might be caused by impact noise was not fully understood until the early 1970s and the judge concluded that the publication of the Ministry of Labour pamphlet *Noise and the Worker* marked the point where a reasonably careful employer ought to have become aware that, if his employees were exposed to a high level of noise, their hearing might be at risk and there were perhaps steps which could and should be taken to eliminate or at least reduce the danger.

Following the hearing in 1983 of a series of actions claiming damages for noise-induced hearing loss sustained whilst working in the ship building industry, it was established that an employer was not negligent at any given time if he followed a recognised practice which had been followed throughout industry for a substantial period, though that practice may not have been without mishap and at that particular time, the consequences of a particular type of risk were regarded as inevitable. Accordingly, 1963 marked the dividing line between a reasonable policy of following the same line of inaction as other employers in the trade and a failure to take positive action. After the publication of *Noise and the Worker*[3] there was no excuse for ignorance.

These cases also confirmed that claimants are only entitled to recover compensation for the additional detriment to their hearing caused during the period when the employers were in breach of their duty – see *Thompson* v. *Smiths Ship Repairers (North Shields) Limited* (1984) 1 All ER 881.

Some 26 years after the publication of *Noise and the Worker*[3], a comprehensive set of regulations was introduced to control the exposure of workers to the effects of noise – The Noise at Work Regulations 1989. These regulations, effective from 1 January 1990, require employers to eliminate or reduce noise exposures above prescribed levels subject to an overriding requirement to reduce, so far as is reasonably practicable, the exposure to noise of employees. What is reasonably practicable will vary according to the circumstances. An employer is required to weigh the quantum of risk against the money, time and trouble involved in remedying the problem and whilst he is not required to incur such cost as to make his business uncompetitive, the protection of the physical health of his employees must demand a high priority. Where it is not possible to reduce noise below the prescribed level protective equipment must be provided and the employee must wear it.

Technological and medical advances in recent years have increased the awareness of the possible relationships between diseases and working environments including contact with injurious substances and operation of machinery. Attempts are constantly being made to extend fields of potential liability. In 1980 a man who developed symptoms of vibration-induced white finger after working as a caulker/rivetter for many years failed in his claim for damages for personal injury against the Ministry of Defence as employers since in 1973 when the complaint arose little was known of the condition. See *Joseph* v. *Ministry of Defence* Court of Appeal Judgement 29 February 1980 – *The Times* 4 March 1980. Since that time knowledge of the condition has increased and vibration-induced white finger acquired in certain occupations has been introduced as a prescribed disease by the Social Security (Industrial Injuries) (Prescribed Diseases) Amendment Regulations 1985 with effect from 1 April 1985. There have now been a few successful claims brought by employees against their employers following development of the condition of vibration-induced white finger. In the case of *Heal* v. *Garringtons*, unreported, 26 May 1982, it was held that a workman had been exposed to excessive levels of vibration produced by a dressing tool used on a pedestal grinder.

In 1996, the Court of Appeal upheld the judgment in the case of *British Coal* v. *Armstrong and Others* (*The Times*, 6 December 1996, CA) which held that, in the light of the evidence, British Coal should have recognised by 1973 that the work undertaken by the claimants gave rise to forseeable risk of vibration white finger (VWF) and that they should have taken effective precautions to guard against the risk by 1976.

It is generally accepted that industry should have been aware of the risk of vibration-induced white finger from certain processes involving exposure of vibration-inducing equipment by 1976. This does not necessarily imply that an employer is liable from that date as the courts have shown a willingness to realise that an employer cannot modify processes overnight. In another unreported case – *Shepherd* v. *Firth Brown* 1985 – the judge allowed three years after the date of knowledge for the employers to modify an engineering process to reduce vibration.

7.3 Public Liability insurance

In addition to the statutory duty to insure against his legal liabilities to his employees an employer will usually insure against his liability to others. This liability may arise from his occupation of premises, the duty to visitors being governed by the Occupier's Liability Act 1957. In addition to the various statutory controls to eliminate the effects of pollution and environmental hazards under the Control of Pollution Act 1974 and the Health and Safety at Work Act 1974, and more recently the Environment Protection Act 1990, the common law has developed doctrines that impose strict liability for the escape of things likely to do damage should they be allowed to escape. See *Rylands* v. *Fletcher* [1861] 73 All ER Reprints No. 1. The occupier may even owe a duty to trespassers in certain circumstances, at least to act with humane consideration. This concept is of particular relevance to injuries to trespassing children, for example see *British Railways Board* v. *Herrington* [1971] 1 All ER 897. The duty an occupier of premises owes to persons other than visitors is now contained in the Occupier's Liability Act 1984. Public Liability insurance has been developed to indemnify the insured against this type of risk, the insurer providing cover against liability for injury to or illness of third parties (other than employees) and loss of or damage to third party property and including claimants' costs and expenses on the same basis as the Employer's Liability policy. It must be emphasised that for Public Liability policies to operate the occurrence must be accidental in origin, for example damage caused to plaster removed by an electrician to facilitate examination of wiring would not be covered. The injury or damage must also occur during the period of insurance and in connection with the business as defined in the policy although it is normally emphasised that the interpretation embraces the insured's legal liability arising from associated activities such as canteens, sports clubs, works fire service, medical facilities and the like.

The Public Liability policy will exclude liability arising out of the ownership, possession or use by or on behalf of the insured of a mechanically propelled vehicle, vessel or craft, the insurances of which are more properly the province of other policies. With regard to motor vehicles liability is often incurred by an employer where the driver of a vehicle who is acting as a servant or agent of the employer is negligent causing injury for which the employer is vicariously liable. However, the insurance against liability in respect of the death of or bodily injury to any person caused by or arising out of the use of a vehicle on a road is compulsory by virtue of the Road Traffic Act 1972, see part VI of the Act – Third Party Liabilities.

It is also not the intention of the insurer to provide cover against the insured's liability for damage to property belonging to or in the custody, possession or control of the insured which is more properly the province of material damage policies although often cover is extended in relation to the personal effects including motor vehicles of employees, but in each case legal liability for such damage must devolve on the employer before the policy cover operates.

One particular area of third party liabilities which has been the subject of recent reform is that relating to the supply of defective products. In July 1985 the EEC adopted a directive on product liability[4] which had to be introduced into the laws of Member States within three years. This was achieved in this country following the implementation of the Consumer Protection Act 1987. This Act creates a new civil liability for injury or damage caused wholly or partly by a defect in a product. The existing legal framework, under which a person could bring a claim for damage resulting from defective goods either by means of an action in contract or tort, is retained.

Prior to the Consumer Protection Act a very limited form of strict liability existed in the form of statutory liability in contract arising from the direct supply of defective products. This is defined by the Sale of Goods Act 1893 as amended by the Supply of Goods (Implied Terms) Act 1973 (now consolidated into the Sale of Goods Act 1979) and the Unfair Contract Terms Act 1977.

The eventual consumer who sustains injury or damage may be able to succeed in an action in tort under the principle enunciated in *Donoghue (McAlister)* v. *Stevenson* [1932] All ER Reprints 1, if he is able to prove not only that a product was defective and it was that which caused the injury or damage but also that the defendant has failed in his duty of care. The defendant may raise various defences to the claim including contributory negligence or a defence based on the 'state of the art' whereby he asserts that he exercised all reasonable care in accordance with the present level of technological knowledge. This defence is also available to defendants in relation to claims brought under the Consumer Protection Act.

The Products Liability Insurance policy is designed to cover this type of risk, indemnifying the insured against his liability for bodily injury or illness to persons or loss of or damage to property caused by products sold, supplied or repaired by the insured although damage to the defective product itself is excluded.

7.4 Investigation, negotiation and the quantum of damage

Once a claim has been intimated by an injured person or by a solicitor on his behalf the insurer undertakes a detailed investigation into the circumstances of any accident prior to taking any decision regarding liability. Even before this stage is reached it is incumbent upon the insured to notify the insurer of any accident which may be the subject of indemnity under the policy. Some cases, for example fatal accidents, are serious enough to warrant immediate investigation to obviate the possibility of alteration or destruction of physical evidence and to ensure that the witnesses' evidence is secured before the facts become clouded through the passage of time. In fatal cases it is usual for an insurer to instruct solicitors to represent the insured at the inquest who will then report on the proceedings and where necessary obtain the depositions.

Any investigation will usually combine observation of the scene of any accident including the examination of any machinery or apparatus involved and the taking of detailed statements from witnesses, independent where possible. If litigation is in prospect full proofs of evidence may be obtained and particular regard paid to the demeanour of the individual in relation to the form and manner in which he is likely to reproduce his oral evidence in the court. Where both sides in an action produce technical and expert reports a judge will decide which opinion he is disposed to accept. Where the accident or loss involves complex machinery or systems of work it is usual for both the defendant's and plaintiff's advisers to use the services of consulting engineers. However, it must be remembered that the rules of the Supreme Court and the rules of the County Court stipulate that where a party intends to rely on expert evidence, the substance of that evidence must be disclosed to the other parties in the form of a written report. In the case of any other oral evidence he shall serve on all other parties written statements of oral evidence he intends to adduce, both within specified time limits. The question of what constitutes the substance of that evidence was dealt with in *Ollett v. Bristol Aerojet Limited* [1979] 3 All ER 544 where the judge confirmed that this phrase embraced both the substance of a factual description of the machine and/or the circumstances of the accident *and* his expert opinion in relation to that accident. Where the employer has himself carried out detailed enquiries into the circumstances of an accident a document containing the results of that enquiry may be discoverable, i.e. it must be revealed to the plaintiff's advisers, if litigation is not the dominant purpose of the raising of the document – see in particular *Waugh v. British Railways Board* [1979] 2 All ER 1169.

It is likely that the recommendations in Lord Woolf's Civil Justice Review which encourage a 'cards on the table' approach will mean that it will be increasingly difficult to claim particular documents are privileged.

With the aid of experts the insurer will assess the evidence and decide whether liability will attach to the insured. A condition in the policy stipulates that the insured themselves must make no admission of liability, even impliedly, without the consent of the insurers. Conversely insurers do not admit liability to a third party on behalf of their insured without prior consent. Repudiation of a claim will only be made after careful consideration of all of the evidence because litigation is both costly and uncertain in outcome.

The next stage is to assess the quantum of damage, in property damage cases often with the aid of loss adjusters and in personal injury cases with the assistance of medical experts. A medical examination will be arranged where the nature of the injury is sufficiently serious to warrant this expense and where possible an exchange of medical evidence with the plaintiff's advisers is undertaken. Once medical evidence has been clarified the insurer will commence negotiations with a view to agreement of any amount to be paid in settlement of the claim.

The law of damages is complex and in a state of constant evolution. Consequently a full discussion and analysis is beyond the scope of this text. As a brief summary, damages may be classified in the following way (for a full analysis see McGregor on Damages[5] – and for up-to-date case law see Kemp[6]):

(1) Pecuniary loss
This may be subdivided into:

(a) Past losses – Included under this heading would be the claimant's net loss of earnings, medical expenses, nursing fees, damage to clothing, cost of repairs to property, all of which must have been reasonably incurred.

  For accidents occurring or where a claim for benefit naming a disease was or is made on or after 1 January 1989 for which damages above £2500 were paid on or after 3 September 1990 the compensator could deduct all relevant State Benefits from the damages and repay them to the Department of Social Security by virtue of the Social Security Act 1989 and the Social Security (Recoupment) Regulations 1990.

  On 6 October 1997 the Social Security (Recovery of Benefits) Act 1997 came into force, replacing all previous recoupment regulations. Under this Act, compensators must now pay to the DSS a sum equivalent to the amount of recoverable State Benefits paid during the relevant period, which is the period between the date of the accident (or in disease cases the date recoverable benefit is first claimed) and the date of settlement.

  The compensator can then reduce the amount of compensation paid in respect of loss of earnings, past care costs and/or past loss of mobility, by way of a direct set-off against amounts payable to the DSS on a like for like basis.

(b) Future losses – The court must attempt to predict the plaintiff's needs and the future costs thereof. If the plaintiff can show that his income will be substantially reduced in the future and this will result directly from the accident then this is a recoverable head of damages. In its simplest form it will be calculated by reference to the plaintiff's future earning capacity in relation to his notional pre-accident earnings and multiplied by the number of years over which the loss will exist, due allowance being made for the contingencies of life – see *Lim Poh Choo v. Camden and Islington Area Health Authority* [1979] 2 All ER 910.

(c) Loss of future earning capacity – Where the plaintiff has a disability but has returned to equally remunerative employment, compensation may be payable for the risk of loss of opportunity to earn in the future – see *Moeliker v. Reyrolle* WLR 4 February 1977.

(d) Loss of profit – In relation to some aspects of this head of damage see *Spartan Steel and Alloys Ltd v. Martin and Company (Contractors) Limited* [1972] 3 All ER 557 and *SCM (UK) Limited v. W.J. Whittle and Son Limited* [1970] 2 All ER 417.

(2) Non-pecuniary losses
Compensation for pain, suffering and loss of amenity falls into this category. This is awarded by way of general damages and the courts do not apportion individual amounts to each subdivision but merely make a global award. The potential value of any claim must be assessed by reference to previous awards falling within the same general category making due allowance for any individual distinguishing characteristics.

There are other heads of damages including loss of expectation of life and in particular the statutory entitlement of dependants of the deceased person under the Fatal Accidents Act 1976.

Any award or negotiated settlement should also take into account any reduction in the damages possible by virtue of the Law Reform (Contributory Negligence) Act where the plaintiff suffers damage partly as a result of his own fault. The criterion for the proportion of assessment is the degree to which the plaintiff has departed from the accepted norm as compared to the degree of culpability attached to the defendant. The statute itself refers to a reduction in damages 'to such extent as the court thinks just and equitable having regard to the claimant's share in the responsibility for the damage'. Contributory negligence is not always easy to establish. In particular, momentary inadvertence by an employee where the employer is in flagrant breach of his statutory duty will not suffice to mitigate damages, for example see *Mullard* v. *Ben Line Steamers Limited* [1971] 2 All ER 424. Although contributory negligence can amount to a significant degree of culpability it cannot equate to 100% – see *Pitts* v. *Hunt and Another* [1990] 3 All ER 344.

7.5 General

The role of the insurer extends beyond the mere limitations of indemnifying an employer against his liability for certain injury or damage. Accident prevention is of benefit to both the insurer and the insured because in the final analysis premiums are influenced by the claims cost ratio. The social benefits of accident prevention are of course impossible to measure in terms of the avoidance of personal suffering and financial loss. The insurers employ experienced surveyors whose job embraces risk reduction in a direct sense through their observation of potential hazards during surveys prior to the arrangements of Employer's Liability, Public Liability and Engineering insurances resulting in the making of recommendations to improve the risk to be insured.

References

1. Rideout, R. W., *Principles of Labour Law* – 5th edn, Sweet & Maxwell, London (1989)
2. *Disability Rights Handbook*, 14th edn, April 1989–1990, The Disability Alliance Educational and Research Association, London (1989) (revised annually). See also Department of Social Security leaflet FB2 – *Which Benefit?* from local Department of Social Security office
3. Department of Employment, Health and Safety at Work booklet No. 25, *Noise and the Worker*, HMSO, London (1974) (first published in 1963 – out of print)
4. EEC, *Directive on the approximation of the laws, regulations and administrative provisions of the Member States concerning liability for defective products*. Directive No. 85/374/EEC, official journal No. 1210/29, Brussels (1985)
5. McGregor, Harvey, *McGregor on Damages*, 15th edn, Sweet & Maxwell, London (1988)
6. Kemp, D. A. M., *Quantum of Damages*, Vol. 2, *Personal Injury Reports*, Sweet & Maxwell, London (1989)

Chapter 8
Civil liability
E. J. Skellett

8.1 The common law and its development

The term 'the common law' means the body of case law of universal, or common, application formed by the judgements of the courts. Each judgement contains the judge's enunciation of the facts, a statement of the law applying to the case and his *ratio decidendi* or legal reasoning for the finding to which he has come. The judgements are recorded in the various series of Law Reports and have thus developed into the body of decided case law which we now have and which continues to develop.

The doctrine of precedent whereby an inferior court is bound to follow the judgement of a higher court ensures consistency in the law. Thus an earlier judgement of the Court of Appeal will bind a High Court or county court judge considering a similar situation and a decision of the House of Lords is binding on all inferior courts although the House itself is free to reappraise its previous judgements.

The common law is not a codified body of law clearly defined in its extent and limits. New law is being made all the time. Judges are asked to adjudicate on sets of circumstances which previously might not have been considered by the courts. Moreover a judge, in applying the established principles of common law to the facts he is considering, might well distinguish that particular case from earlier decided cases. Finally in determining whether in a case there has been compliance with standards such as that of 'reasonable care' the judge will of necessity approach the problem in the light of contemporary knowledge and thinking. Thus what is adjudged reasonable conduct in 1950, say, will not necessarily be adjudged reasonable in 1980. In these ways, judges bring up to date the body of common law and adapt and develop it in accordance with the standards and social principles of the era. Such changes are of course slow and gradual, but the common law is also subject to more drastic and immediate change by Parliament, examples being the Employer's Liability (Defective Equipment) Act 1969 and the Occupier's Liability Acts 1957 and 1984. Although Parliament thus exercises dominance over the common law, the statutes in their turn are interpreted by the judges following legal rules and principles already well established.

8.2 The law of tort

This concerns the legal relationships between parties generally in the everyday course of their affairs, the duties owed one to the other and the legal effect of a wrongful act of one party causing harm to the person, property, reputation or economic interests of another.

The law of tort covers relationships generally, compared with the law of contract which applies where two or more parties have entered into a specific relationship between themselves for a specific purpose.

Three separate branches of the law of tort are trespass, nuisance and negligence, the latter being by far the most important and applying in particular to the field of an employer's liability for accidental injury to his employee.

8.2.1 Trespass

This is the oldest branch of the law of tort. An action for trespass is nowadays generally confined to the intentional invasion of a man's person, land or goods involving, for example, such civil claims for damages as those resulting from battery, assault, false imprisonment, unlawful entry onto the land of another. In the latter case, apart from legal action, direct action can be taken against the trespasser using reasonable force to regain possession against, for example, squatters or 'sit-in' demonstrators. It also includes claims for conversion, an intentional dealing with a chattel constituting a serious infringement of the plaintiff's right of possession.

8.2.2 Nuisance

There are two forms, private nuisance or public nuisance. An action for private nuisance lies only where there has been interference with the enjoyment of land and is appropriate where an occupier of land has acted in such a way as to harm his neighbour's enjoyment of his land. It need not be a deliberate interference and includes such cases as the emission of smoke, fumes or excessive noise. The interference must be sufficiently significant and must be unreasonable. In deciding if it is, the court will take into account all circumstances including the reason for the alleged nuisance, the locality (e.g. whether rural or industrial), the ordinary use of the land and the impracticability of preventing the nuisance.

The second classification of nuisance, public nuisance, constitutes a criminal offence as well as being an actionable wrong at civil law for which damages may be claimed for any injury or damage caused. Public nuisance relates to acts interfering with the public at large and includes, for example, obstruction of the highway, leaving open a cellar flap or leaving unlit scaffolding abutting onto the highway.

8.2.3 Negligence

A broad definition is careless conduct causing damage or injury to another.

Actions based upon the tort of negligence are far commoner than those based upon other torts. Distinctions are not exclusive. Very often the same facts can found an action both in negligence and nuisance. There are three elements necessary to establish a case in negligence:

1 that there is a duty of care owed by one party to the other,
2 that there has been a breach of that duty,
3 that the breach of duty has resulted in damage.

8.2.3.1 The duty of care

To whom is this owed? In the case of *Donoghue* v. *Stevenson*[1] this was defined as follows:

> 'You must take reasonable care to avoid acts or omissions which you can reasonably foresee would be likely to injure your neighbour.'

Neighbours are defined as:

> 'Persons who are so closely and directly affected by my act that I ought reasonably to have them in contemplation as being so affected when I am directing my mind to the acts or omissions which are called in question.'

There are no hard and fast rules as to who might or might not fall into this category, and this must be examined in each case. In some situations, the public at large may be owed a duty, for example by a motorist. In others, a duty is more closely defined. An employer owes a duty of care in tort to his employee, a manufacturer to the consumer, a solicitor to his client.

The standard of care owed
This requires an examination of the facts of the particular circumstances. The magnitude of the risk of injury and the gravity of the consequences of an accident must be weighed against the cost and difficulty of obviating the risk. A considered decision has to be made. Even though a risk may not warrant extensive precautions, the particular process, place or person may have features that make these vital. In *Paris* v. *Stepney BC*[2], for example, the House of Lords held goggles should have been provided for a one-eyed man doing work where there was a risk of metal particles striking the eye although the risk of this happening was such that for a man with normal sight it could be ignored. The question is put succinctly by Denning LJ in *Latimer* v. *AEC Ltd*[3]:

> 'It is a matter of balancing the risk against the measures necessary to eliminate it.'

The New Zealand courts give a convenient and simple approach to the issue in the case of *Fletcher Construction Co. Ltd* v. *Webster*[4]:

1 What dangers should the defendant, exercising reasonable foresight, have foreseen?
2 Of what remedies, applying reasonable care and ordinary knowledge, should he have known?
3 Was the remedy, of which he should have known, for the danger he should have foreseen, one he was entitled to reject as unreasonably expensive or troublesome?

8.2.3.2 Breach of duty

Once the existence of the duty of care which arises from the relationship of the parties concerned and its standard are established, one has to consider whether or not there has been a breach of that duty, and if so consideration can be given to the next question.

8.2.3.3 Res ipsa loquitur

This Latin maxim means literally 'the thing speaks for itself'. In other words the circumstances of the accident giving rise to the action are such as impute negligence on the part of the defendant, being an event which, if the defendant had properly ordered his affairs, would not have happened. If this plea by a plaintiff is accepted by the court then a presumption of negligence is raised against the defendants. In other words, effectively it is for the defendant to prove the absence of fault rather than for the plaintiff to prove fault. The defendant can set aside the presumption against him by:

1 Proof of reasonable care having been taken.
2 An alternative explanation for the accident which is equally probable and which does not involve negligence on the part of the defendant.
3 A complete analysis of the facts, i.e. the defendant laying before the court all the facts of the case and inviting full consideration of liability.

Illustrations of the application of this maxim are such cases as bricks falling from a bridge onto a person walking underneath or cargo falling from a crane onto an innocent passerby, i.e. where one would say that prima facie the accident could not have happened without someone's fault.

8.2.3.4 The resultant damage

The damage must result from the negligent act or omission and be caused by it. In other words it must be a direct consequence. Most cases of injury are straightforward but sometimes unexpected complications arise, as in the case of *Smith* v. *Leach Brain & Co. Ltd*[5] where a plaintiff was entitled to recover damages for cancer developing from a burn on the lip caused by molten metal. This was a direct result of the burn. However, the chain of causation must not be broken – there must not be a *novus actus interveniens*, i.e. an act of another party intervening between the defendant's breach and the loss, or a *nova causa*, i.e. an independent and unforeseeable cause intervening. For example, in *McKew* v. *Holland and Hannen and Cubitts Ltd*[6] it was held that a workman who had sprained his ankle and later fell down stairs when the ankle gave way, resulting in his breaking his leg, could recover from the original wrongdoer damages for the ankle injury but not for the fractured leg because he himself had been negligent for not holding on to the handrail. His negligence was held to constitute a *novus actus*.

If there are more than one possible causes of an injury, it is for the plaintiff to prove causation – *Wilsher* v. *Essex Health Authority*[7]. However, where a pedestrian was injured by one car then further injured by being thrown into the path of a second, it being impossible to say what proportion of injury was caused by each motorist, it was held that the plaintiff did not have to go so far as to prove the extent of injury caused by each – *Fitzgerald* v. *Lane*[8].

8.3 Occupier's Liability Acts 1957 and 1984

The 1957 Act defines the duty owed by the occupiers of premises to all persons lawfully on the premises in respect of:

> 'Dangers due to the state of the premises or to things done or omitted to be done on them. Section 1(i).'

The liability is not confined to buildings and has been held to include, for example, that of the main contractors retaining general control over a tunnel being constructed – *Bunker* v. *Charles Brand & Son Limited*[9].

Section 2 defines the standard of care, owed by the occupier to the persons lawfully on the premises, namely:

> 'A common duty of care to see a visitor will be reasonably safe in using the premises.'

Then by s. 2(3) 'A person present in the pursuance of his calling may be expected to appreciate and guard against any special risks ordinarily incidental to it, so far as the occupier leaves him free to do so'. In other words this class of visitor is expected to use his own specialist knowledge.

Under s. 2(4) 'A warning or notice does not, in itself, absolve the occupier from liability, unless in all the circumstances it was sufficient to enable the visitor to be reasonably safe'. Whilst the occupier could, under this section, avoid his liability by a suitably worded notice, this is superseded by the Unfair Contract Terms Act 1977, which provides that it is not permissible to exclude liability for death or injury due to negligence, by a contract or by a notice and this applies to a notice under s. 2(4) of the Occupier's Liability Act 1957. The 1957 Act made no provision for those outside this category of lawful visitors, i.e. contractors, invitees and licensees. The 1984 Act extended the classes of persons to whom the duty of care is owed to those exercising public and private rights of way, ramblers and trespassers. In the latter case the Act was directed to alleviate the position of the innocent, such as the young child or someone walking blithely unaware he had no right to be there, rather than the deliberate trespasser.

8.4 Supply of goods

In the normal course of obtaining goods, the purchaser can reasonably expect to be supplied with goods that are fit for the purpose for which he purchased them.

8.4.1 Manufacturers

They owe a duty of care to the consumer of their products independently of any rights the purchaser of their products may have under contract law, against the supplier to them of goods. Thus a consumer may be able to sue both his supplier and the manufacturer.

The leading case is *Donoghue* v. *Stevenson*[1] which established the principle, the House of Lords holding that someone who drank ginger beer from an opaque bottle, given her by a friend, and who became ill from the presence of a snail in the bottle was entitled to damages from the manufacturers if she could prove her case.

The manufacturer's duty is to take reasonable care in manufacture to ensure that the product is without defect and not liable to cause injury.

There is no liability on a manufacturer if there is the opportunity of intermediate examination particularly where this is expected, which it could not be in the case of a sealed opaque bottle. Nor for instance is a manufacturer liable to a workman injured by using defective goods the manufacturer supplied which an employer examines, sees are defective but decides to keep in use albeit only until they can be replaced.

8.4.2 Consumer Protection Act 1987

By s. 2, where damage is caused wholly or partially by a defect in a product, then producers, own-branders, importers and suppliers are liable for that damage.

Anyone damaged by a defective product has a right of action against those from whom they obtained the finished product or those involved in the production process. The Act does not cover liability for economic loss (even though recognised by the common law in *Junior Books Co. Ltd* v. *Veitchi*[10]) or damages below £275 or claims against repairers and second-hand dealers. Liability is non-excludable by contract, notice or otherwise.

Civil liability

The Act specifically makes it a defence that the product was supplied other than by way of the defendant's business, e.g. by gift. It also provides for a 'development risks' defence, i.e. that the defect was not one the defendant was aware of at the time, given the state of the scientific and technical knowledge then prevailing.

8.5 Employer's liability

An overall statement of the duty owed by an employer to his employees is that he must take such care as is reasonable for the safety of his employees. That duty is owned to each and every employee as an individual, taking into account his own weaknesses and strengths, and is owed wherever the employee may be in the course of his employment, on or off the employer's premises. It is a duty which the employer owes personally to the employee and the employer remains responsible for a breach of that duty even if he has delegated the performance of that duty to someone else, for example to a safety consultant who might have a separate liability. The same applies if he has put his employee to work under the order of another party – *McDermid* v. *Nash Dredging and Reclamation Co. Ltd*[11].

The employer can be held liable either directly for breach of his own duties or vicariously. Vicarious liability arises where an employee or an agent of the employer has acted negligently and caused injury to another employee. The employer is legally liable for the wrongful act or omission where it has been performed in his interests. However, he is not liable if the employee acts negligently on a frolic of his own independently of his employment. *Smith* v. *Crossley Bros Ltd*[12] illustrates this, where, as a joke, two apprentices injected compressed air into the body of a third and the employers were held not liable.

The employer's duty at common law can conveniently be considered under five heads. Obviously each will turn on the particular circumstances involving one or more of these elements and it is impossible to give more than general guidelines. The heads are:

1 system of work,
2 place of work,
3 plant and equipment,
4 supervision and/or instruction,
5 care in selection of fellow employees.

8.5.1 System of work

The employer is obliged to set up and operate a safe system of work, and it is a question of fact in every case what is safe. This includes such matters as the co-ordination of activities, the layout and arrangement of the way a job is to be done, the use of a particular method of doing a job. The employer is expected to plan and draw up an original method of operation which is safe and free, so far as possible, from foreseeable cause of injury. Regard will be held to established practice and absence of accident in assessing what is safe, but the court will still examine the practice to decide if it is safe. In *General Cleaning Contractors Ltd* v. *Christmas*[13] Lord Oakley said in his judgement:

> 'the common law demands that employers should take reasonable care to lay down a reasonably safe system of work.'

He continued that workmen even though experienced and competent to lay down a system themselves should not be expected to do so, making their decisions at their workplace where the dangers are obscured by repetition, compared with the employer who performs his duty in the calm atmosphere of a boardroom with the advice of experts.

8.5.2 Place of work

The employer is under a duty at common law to provide a reasonably safe place of work, relating to such matters as the provision of gangways clearly marked and free of obstruction, and the maintenance of floors and staircases. The duty is fulfilled through regular inspection of the workplace and keeping it in a safe state, free of hazard so far as reasonably practicable. It does not extend to protection from abnormal hazards which the employer could not reasonably have foreseen. For example, whilst in conditions of ice and snow, paths must as far as possible be sanded before the normal time for employees to arrive at the premises, if there is a sudden totally unexpected snowfall, the employer is not liable if paths are slippery or obstructed until he has had reasonable opportunity to remedy the situation.

The duty extends to any place at which the employee works whether belonging to his employer or not, but it will depend on the circumstances whether the employer should have inspected them before sending his employees to work there, and perhaps had steps taken to make them safer. For example, no court would suggest the employer of a plumber sent out to work at a private house should first send the foreman or supervisor to inspect the house unless the employer had prior knowledge of some particular feature of the premises which introduced added risk. In most cases involving factory or site accidents the relevant section of the Workplace Regulations 1992 or the Construction (Health, Safety and Welfare) Regulations 1996 will be pleaded in addition to the duty at common law.

8.5.3 Plant and equipment

The employer owes a duty to his employee to provide safe and proper plant and equipment which must also be suitable for the purpose to which it is put.

It is a far-ranging aspect of the employer's duty. In the first place the employer may have failed completely to provide equipment necessary for the safe performance of work, for example mechanical lifting equipment for a load too heavy to be manhandled.

Equipment supplied may be unsuitable for the particular function, or it may be the proper equipment but inadequately maintained or defective.

Consideration will be given in deciding if the employer is liable to the procedure followed for reporting and rectifying defects, routine maintenance, the issue of small items of plant and such like.

This aspect is relevant also to the question of whether an employer has provided protective equipment such as gloves, goggles and ear-muffs to reduce or prevent exposure to foreseeable risk of injury.

Where a claim for damages arises out of an accident in a factory, the appropriate sections of PUWER will be relied upon, for example relating to the guarding of machinery, in addition to the duty at common law.

The Employer's Liability (Defective Equipment) Act 1969 discussed later is relevant to this aspect too.

8.5.4 Supervision and/or instruction

An employer must take such care as is reasonable to ensure adequate and proper supervision over and instruction to his employees. What is reasonable must depend on the circumstances, including the complexity of the work to be done, the technicality of the equipment concerned and the age and experience of the workman. It must be obvious that if a young inexperienced man is set to work on a complicated machine or a complicated task where he can injure himself the employer will be held liable. It must not be thought, however, that an employer can leave even a senior experienced man to his own devices. Supervision and instructions are a matter of degree but always the courts will impute to the employer a superior knowledge of the dangers and risks in a work system with the consequent duty to supervise and instruct his employees.

8.5.5 Care in selection of fellow employees

This aspect of an employer's duty is of less significance since the employer will be held vicariously liable for the act of an employee who negligently injures another, which was not always so.

It is most relevant to the type of case where an employee indulging in horseplay or fighting has injured another and the man concerned has a history of such activities to the knowledge of the employer, who has taken no steps to dismiss him or prevent a recurrence.

8.6 Employer's Liability (Defective Equipment) Act 1969

Prior to the passing of this Act where a workman sued his employer in respect of injury caused by a defective tool or item of plant supplied by the employer to the employee it was a defence for the employer to prove that he did not know and could not reasonably have known of the defect and that he had exercised reasonable care when he obtained the item concerned, by going to a reputable manufacturer or supplier. This was the rule in *Davie* v. *New Merton Board Mills Limited*[14]. The Act changed the law and imposed liability on the employer where an employee was injured in consequence of a defect in equipment provided by his employer for the purpose of the employer's business, if the defect was attributable wholly or partly to the fault of a third party (whether identified or not). In other words the employer no longer has a defence if he provides defective equipment to his employee which results in injury and the defect was the fault of another party. This does not mean that the employer is without remedy against that other party. He is entitled to bring an action against the supplier in respect of the defective plant, but must be able legally to prove his case against the supplier. It is perhaps unnecessary to add that an employer is liable irrespective of the Act if it can be proved that the defect should have been found by the employer on inspection before being put into use or if an employer had caused or permitted his employee to keep in use defective items of plant.

8.7 Health and Safety at Work etc. Act 1974

Although ss. 2–8 of the Act impose general duties on parties including employers, failure to comply with the obligations imposed by the Act itself does not provide grounds for a civil claim. However, section 47(2) of the Act stipulates that an action can be based on a breach of Regulations made under the Act, unless the Regulation has a specific exclusion.

8.8 Defences to a civil liability claim

The first and obvious defence which may be raised is a denial of liability which may be based on a variety of grounds.

1 That the duty alleged to have been breached by the defendant was never imposed on him in the first place, for example in an employee's claim against his employer that the plaintiff was not an employee but was working for another company.
2 That the nature of the duty was different from that pleaded against the defendant.
3 That the duty owed was complied with and not breached.
4 That the breach of duty did not lead to the damage.
5 That the plaintiff was himself guilty of contributory negligence resulting wholly in the damage.

Secondly in the defence it may be pleaded that conduct of the plaintiff, constituting contributory negligence, caused and/or resulted in part in the damage he suffered and that any damages which might be payable to him should be reduced accordingly – the Law Reform (Contributory Negligence) Act 1945. By way of example, that he failed to see a hole into which he fell. Obviously such a consideration only comes into play in the event of a finding that the defendant is liable. The court will then assess the respective blameworthiness of the parties to decide whether there are grounds for finding the plaintiff partly to blame and, if contributory negligence is established, the court will determine the amount of damages the plaintiff would receive if he succeeded in full and then discount these by the proportion to which the plaintiff is himself found to blame.

Thirdly there is the situation where the accident is the fault not of the defendant sued but of some other party. If another party is blamed in the defence, the usual result is that they are joined in as a co-defendant by the plaintiff, and he sues both. However, if a defendant considers that if he is liable to the plaintiff, then he in turn is entitled to recover from someone else any damages he has to pay to the plaintiff in which case that person can be joined in the proceedings by the defendant as a third party. An example of the circumstances where there may be third party proceedings is one where an injured workman who has fallen into a hole at the place where he works sues his employer, who then brings in by third party proceedings the contractor who had left the hole unfenced. In such a case, the plaintiff would have to establish that the defendant was liable to him, and in turn the defendant would then have to prove his case against the third party. The third party will be liable only if the defendant is liable to the plaintiff.

Compare this situation with an action where the plaintiff sues more than one defendant, such as, in the example given above, both suing his employer and the contractor direct. The plaintiff might fail against both defendants, succeed against one or the other or succeed against both, the judge apportioning the degree of liability attaching to each defendant. In the case of *Fitzgerald* v. *Lane* the House of Lords held that where there were two or more defendants, the first consideration was whether the plaintiff had proved his case against the defendants, then the question of whether he was himself negligent and his damages should be reduced accordingly, and finally the apportionment of liability between the defendants themselves.

Thus a pedestrian who ran into the road and was hit by a car and then by another and who was held equally to blame with the car drivers, had his damages reduced by 50% on account of his own negligence. The car drivers' 50% share of the blame was then apportioned between them at 25% each.

8.8.1 Joint tortfeasors

Where two or more parties are responsible for breaches of duty leading to a single injury, i.e. the same damage, they are jointly liable as wrongdoers, in whatever proportion of fault is determined from the circumstances. The simplest example is where two vehicles collide, due to the fault of both drivers, injuring an innocent passenger. The passenger's claim may be enforced against either tortfeasor, who can, under the Civil Liability (Contribution) Act 1978, then claim contribution from the other to the extent of the other's liability.

Another illustration is where both the employer and another contractor engaged on work at the same building site are jointly liable for injury to employee. It must be noted that generally, an employer is not liable for the torts of an independent contractor, unless the work to be carried entails particular danger. Furthermore, an employer cannot get out of his liability for his employee's safety by delegating this to a contractor.

8.9 Volenti non fit injuria

Where a person has agreed either expressly or by implication to accept the risk of injury, he cannot recover damages for damage caused to him by that risk.

For this defence to succeed the person concerned must have had full knowledge of the nature and extent of the risk to be run and have accepted that risk of his own free will. Such a defence is available only in extremely limited circumstances in an action by an employee against his employer. In the case of *Smith* v. *Baker*[15] it was pleaded against an employee drilling rock in a cutting over whose head a crane lifted stones. The court held that although he knew of the danger and continued at work he had not voluntarily undertaken the risk of injury from a stone falling from the crane and hitting him.

Such a defence does not apply to an action for damages brought by a rescuer deliberately running risks to rescue someone who has been injured by dangers created by another. The case of *Baker* v. *T.E. Hopkins & Sons Limited*[16] confirms the entitlement of a rescuer to damages for injury in respect of that negligence.

8.10 Limitation

The Limitation Act 1980 stipulates that an action founded on tort shall not be brought after six years from the date when the cause of action accrued but an action for damages for personal injuries or death must be commenced within three years. Otherwise the actions are barred by the statute and the defendant can plead this as a defence. The three years start to run from the date of the accident or date of the plaintiff's knowledge if later. If the injured person dies within the three years, the period starts to run again from date of death or of his personal representatives' knowledge. The saving provisions of 'knowledge' are aimed primarily at the industrial disease cases where the accidental exposure almost invariably dates back many years before the effects of that exposure had developed and were known. There is a careful definition of what is meant by 'knowledge' in s. 11 of the Act.

The Act also permits an overriding discretion to the court to let in late claims where it is equitable or fair to do so. Furthermore in the case of someone under a disability, e.g. an infant or person of unsound mind, the three years do not start to run until the age of 18 or recovery.

8.11 Assessment of damages

Once liability is established the question for consideration is the amount of damages or compensation to be awarded. The object is to put the injured party as far as possible in the same position as before.

In an action for breach of contract or for debt, this amount will already have been defined in the dealing between the parties and is known legally as a liquidated claim. However, in an action for damages in respect of tort, where damage and/or injury has been caused, damages are called unliquidated, i.e. they will have to be calculated and assessed after the event giving rise to the claim.

These damages will comprise special damage and general damages.

8.11.1 Special damage

Special damage consists of heads of specific expenditure or loss as a result of the accident, damaged goods or loss of wages during time off work. In actions for personal injury, it consists primarily of the loss of wages and the figure recoverable is the net wage lost after deduction of income tax and national insurance contributions, i.e. the actual amount the plaintiff would have received in his pocket. He will be awarded both his total loss of wages during total incapacity from work and partial loss if by reason of continuing disability, as a result of the accident, he cannot do his full work or has to change to a lighter job and is thereby earning less. Credit is given for non-contributory payments by the employer such as sick pay. There are also offset against the loss of earnings claim, any tax refunds and unemployment benefit if, after having been certified fit to return to work following an accident, a man cannot return to his old job and is unable to get another. Redundancy payments under the Redundancy Payments Act 1965 as amended are deductible if attributable to the injury. In actions for damages for personal injury the Social Security (Recovery of Benefits) Act 1997 compels, a person making a compensation payment in consequence of an accident to obtain a Certificate of Total Benefit paid by the Department of Social Security and then to deduct from the compensation payment the amount of the benefit, accounting to the DSS for this.

8.11.2 General damages

General damages are those recovered to compensate for pain, suffering and loss of amenity resulting from an injury. Whilst there are no set tariffs there are published guidelines to assist in the assessment and ensure compatibility between awards made by judges and lawyers who also take into account decided cases involving similar injury. However, in calculating the appropriate sum to award, account is taken of such matters as the particular idiosyncrasies of the plaintiff, his age, occupation, hobbies and such like. The court also has regard to the effect of inflation on past awards or similar injuries. For example, the loss of a finger would attract higher damages for an employee who in his spare time was a skilled musician; similarly damages for an incapacitating leg injury to a keen and energetic sportsman would be higher than those for someone in a sedentary occupation with no active hobbies.

Where there is partial or complete incapacity for work continuing after the trial general damages also include a capital sum awarded for future loss of wages. A sum will, where appropriate, be awarded too for loss of opportunity on the labour market. This is intended to compensate for a permanent disability which a prospective employer may take into account in deciding whether to offer employment, compared with a candidate of equal competence who has no such disability.

Awards of damages generally are once and for all. However, there is an exception.

8.11.3 Provisional damages

Section 6 of the Administration of Justice Act 1982 introduced provisional damages for cases where there is a chance that some serious disease or serious deterioration in the plaintiff's condition will accrue at a later date. Appropriate cases include industrial disease claims where there may be a risk of the development of cancer or a malignant tumour in the future. Provisional damages are assessed ignoring that possibility. If it occurs then a further award may be made.

8.12 Fatal accidents

A cause of action in tort, save for defamation by or against a person, survives for the benefit of or to the detriment of the estate under the Law Reform (Miscellaneous Provisions) Act 1934.

On behalf of the estate, loss of earnings to date of death and general damages for pain and suffering during lifetime are claimable, without reference to any loss or gain to the estate resulting from the death.

Under the Fatal Accidents Act 1976 damages for loss of financial support can be claimed by or for the dependants. The definition of dependant is set out in the Act as amended by the Administration of Justice Act 1982 and includes spouse or former spouse, ascendants and descendants as well as adopted children and anyone living with the deceased as spouse, the latter subject to certain conditions.

Damages are calculated by the measure of actual financial loss. Thus the deceased's earnings will be established and the proportion expended on the dependant determined. This will then be multiplied by a number of years' purchase to allow for the length of time the deceased would have worked. A deduction will be made for capitalisation.

The Administration of Justice Act 1982 also introduced a claim for 'bereavement damages' under which a fixed sum is payable by way of damages – the amount is currently £7500 – for loss of a spouse and to parents for the loss of a child.

8.13 'No fault' liability system

Over the years the possibility of compensation being paid to victims of accidents irrespective of responsibility has been discussed and canvassed but not adopted. The attraction of such a system lies in the removal of the conflict between employer and employee over liability for the payment of damages and the consequent expense in time spent by the employer in detailed assessment of fault and in costs. However, such schemes still lead to dispute over the entitlement to compensation or the amount to be paid, such as those cases fought to establish entitlement to payment under the Workmen's Compensation Acts of the 1940s, later repealed. Many points of question remain to be answered, such as, how should such a scheme be funded? By the State, or by privately arranged insurance cover? Would it be practicable? Where would the limits be drawn, both as to the recipients of compensation and the nature of the compensation – damages for injury alone or including income loss? Injury caused solely by accident or including industrial disease and conditions due to the environment? Direct employees only or contractors too? What about road traffic casualties? If they too are included, is this not unfair to the victims of other accidents, such as those in the home?

References (cases referred to)

1. Donoghue *v.* Stevenson (1932) AC 562
2. Paris *v.* Stepney Borough Council (1951) AC 367
3. Latimer *v.* AEC Limited (1953) 2 All ER 449
4. Fletcher Construction Co. Ltd *v.* Webster (1948) NZLR 514
5. Smith *v.* Leach Brain & Co. Ltd (1962) 2 WLR 148
6. McKew *v.* Holland and Hannen and Cubitts Ltd (1969) 2 All ER 1621
7. Wilsher *v.* Essex Health Authority (1989) 2 WLR 557
8. Fitzgerald *v.* Lane (1988) 3 WLR 356
9. Bunker *v.* Charles Brand & Son Limited (1969) 2 All ER 59
10. Junior Books Co. Ltd *v.* Veitchi (1983) AC 520
11. McDermid *v.* Nash Dredging and Reclamation Co. Ltd (1987) 3 WLR 212
12. Smith *v.* Crossley Bros. Ltd (1951) 95 Sol. Jo. 655
13. General Cleaning Contractors Ltd *v.* Christmas (1953) AC 180
14. Davie *v.* New Merton Board Mills Ltd (1959) 1 All ER 67
15. Smith *v.* Baker (1891) AC 325
16. Baker *v.* T.E. Hopkins & Sons Ltd (1959) 1 WLR 966

Further reading

Munkman, J., *Employer's Liability at Common Law*, 11th edn, Butterworth, London (1990)
Heuston, R.F.V. and Chambers, R.S., *Salmond on the Law of Torts*, 18th edn, Sweet & Maxwell, London (1981)
Kemp, D., *Damages for personal injury and death*, Oyez Publishing, London (1980)
McGregor, Harvey, *McGregor on Damages*, 15th edn, Sweet and Maxwell, London (1988)

Appendix 1

The Institution of Occupational Safety and Health

The Institution of Occupational Safety and Health (IOSH) is the leading professional body in the United Kingdom concerned with matters of workplace safety and health. Its growth in recent years reflects the increasing importance attached by employers to safety and health for all at work and for those affected by work activities. The Institution provides a focal point for practitioners in the setting of professional standards, their career development and for the exchange of technical experiences, opinions and views.

Increasingly employers are demanding a high level of professional competence in their safety and health advisers, calling for them to hold recognised qualifications and have a wide range of technical expertise. These are evidenced by Corporate Membership of the Institution for which proof of a satisfactory level of academic knowledge of the subject reinforced by a number of years of practical experience in the field is required.

Recognised academic qualifications are an accredited degree in occupational safety and health or the Diploma Part 2 in Occupational Safety and Health issued by the National Examination Board in Occupational Safety and Health (NEBOSH). For those assisting highly qualified OSH professionals, or dealing with routine matters in low risk sectors, a Technician Safety Practitioner (SP) qualification may be appropriate. For this, the NEBOSH Diploma Part 1 would be an appropriate qualification.

Further details of membership may be obtained from the Institution.

Appendix 2

Reading for Part I of the NEBOSH Diploma examination

The following is suggested as reading matter relevant to Part 1 of the NEBOSH Diploma examination. It should be complemented by other study.

Module 1A:	The management of risk	Chapters	2.1–all 2.2–paras. 8–11 2.3–all 2.4–paras. 1–3 3.8–paras. 1–6 4.7–para. 11
Module 1B:	Legal and organisational factors	Chapters	1.1–all 1.2–all 1.3–paras. 1–6 1.7–para. 2 1.8–all 2.2–paras. 13 and 14 2.6–paras. 1–4
Module 1C:	The workplace	Chapters	1.7–para. 2 3.6–all 3.7–all 4.2–all 4.4–paras. 1–8 4.6–paras. 2 and 4 4.7–paras. 1, 2, 7 and 11
Module 1D:	Work equipment	Chapters	4.3–all 4.4–all 4.5–all
Module 1E:	Agents	Chapters	3.1–all 3.2–all 3.3–all 3.5–paras. 1–6 3.6–all 3.8–paras. 4–7 4.7–paras. 1–4
Module 1CS:	Common skills	Chapter	2.5–para. 7

Additional information in summary form is available in *Health and Safety ... in brief* by John Ridley published by Butterworth-Heinemann, Oxford (1998).

Appendix 3
List of abbreviations

ABI	Association of British Insurers
AC	Appeal Court
ac	Alternating current
ACAS	Advisory, Conciliation and Arbitration Service
ACGIH	American Conference of Governmental Industrial Hygienists
ACoP	Approved Code of Practice
ACTS	Advisory Committee on Toxic Substances
ADS	Approved dosimetry service
AFFF	Aqueous film forming foam
AIDS	Acquired immune deficiency syndrome
ALA	Amino laevulinic acid
All ER	All England Law Reports
APAU	Accident Prevention Advisory Unit
APC	Air pollution control
BATNEEC	Best available technique not entailing excessive costs
BLEVE	Boiling liquid expanding vapour explosion
BOD	Biological oxygen demand
BPEO	Best practicable environmental option
Bq	Becquerel
BS	British standard
BSE	Bovine spongiform encephalopathy
BSI	British Standards Institution
CBI	Confederation of British Industries
cd	Candela
CD	Consultative document
CDG	The Carriage of Dangerous Goods by Road Regulations 1996
CDG-CPL	The Carriage of Dangerous Goods by Road (Classification, Packaging and Labelling) and Use of Transportable Pressure Receptacle Regulations 1996
CDM	The Construction (Design and Management) Regulations 1994
CEC	Commission of the European Communities

CEN   European Committee for Standardization of mechanical items
CENELEC   European Committee for Standardisation of electrical items
CET   Corrected effective temperature
CFC   Chlorofluorocarbons
CHASE   Complete Health and Safety Evaluation
CHAZOP   Computerised hazard and operability study
CHIP 2   The Chemical (Hazard Information and Packaging for Supply) Regulations 1994
Ci   Curie
CIA   Chemical Industries Association
CIMAH   The Control of Industrial Major Accident Hazards Regulations 1984
CJD   Creutzfeldt–Jacob disease
COD   Chemical oxygen demand
COMAH   The Control of Major Accident Hazards Regulations (proposed)
COREPER   Committee of Permanent Representatives (to the EU)
COSHH   The Control of Substances Hazardous to Health Regulations 1994
CPA   Consumer Protection Act 1987
CTD   Cumulative trauma disorder
CTE   Centre tapped to earth (of 110 V electrical supply)
CWC   Chemical Weapons Convention

dB   Decibel
dBA   'A' weighted decibel
dc   Direct current
DETR   Department of the Environment, Transport and the Regions
DG   Directorate General
DNA   Deoxyribonucleic acid
DO   Dangerous occurrence
DSE(R)   The Health and Safety (Display Screen Equipment) Regulations 1992
DSS   Department of Social Services
DTI   Department of Trade and Industry

EA   Environmental Agency
EAT   Employment Appeals Tribunal
ECJ   European Courts of Justice
EC   European Community
EEA   European Economic Association
EEC   European Economic Community
EcoSoC   Economic and Social Committee
EHRR   European Human Rights Report

EINECS	European inventory of existing commercial chemical substances
ELF	Extremely low frequency
ELINCS	European list of notified chemical substances
EMAS	Employment Medical Advisory Service
EN	European normalised standard
EP	European Parliament
EPA	Environmental Protection Act 1990
ERA	Employment Rights Act 1996
ESR	Essential safety requirement
EU	European Union
eV	Electronvolt
EWA	The Electricity at Work Regulations 1989
FA	Factories Act 1961
FAFR	Fatal accident frequency rate
FMEA	Failure modes and effects analysis
FPA	Fire Precautions Act 1971
FSLCM	Functional safety life cycle management
FTA	Fault tree analysis
GEMS	Generic error modelling system
Gy	Gray
HAVS	Hand-arm vibration syndrome
HAZAN	Hazard analysis study
HAZCHEM	Hazardous chemical warning signs
HAZOP	Hazard and operability study
hfl	Highly flammable liquid
HIV+ve	Human immune deficiency virus positive
HL	House of Lords
HMIP	Her Majesty's Inspectorate of Pollution
HSC	The Health and Safety Commission
HSE	The Health and Safety Executive
HSI	Heat stress index
HSW	The Health and Safety at Work, etc. Act 1974
Hz	Hertz
IAC	Industry Advisory Committee
IBC	Intermediate bulk container
ICRP	International Commission on Radiological Protection
IEC	International Electrotechnical Committee (International electrical standards)
IEE	Institution of Electrical Engineers
IOSH	Institution of Occupational Safety and Health
IPC	Integrated polluton control
IQ	Intelligence quotient
IRLR	Industrial relations law report
ISO	International Standards Organisation
ISRS	International Safety Rating System

JHA	Job hazard analysis
JP	Justice of the Peace
JSA	Job Safety Analysis
KB	King's Bench
KISS	Keep it short and simple
LA	Local Authority
LEL	Lower explosive limit
$L_{EP.d}$	Daily personal noise exposure
LEV	Local exhaust ventilation
LJ	Lord Justice
LOLER	Lifting Operations and Lifting Equipment Regulations 1998
LPG	Liquefied petroleum gas
LR	Lifts Regulations 1997
lv/hv	Low volume high velocity (extract system)
mcb	Miniature circuit breaker
MEL	Maximum exposure limit
MHOR	The Manual Handling Operations Regulations 1992
MHSW	The Management of Health and Safety at Work Regulations 1992
MOSAR	Method organised for systematic analysis of risk
MPL	Maximum potential loss
M.R.	Master of the Rolls
NC	Noise criteria (curves)
NDT	Non-destructive testing
NEBOSH	National Examination Board in Occupational Safety and Health
NI	Northern Ireland Law Report
NIHH	The Notification of Installations Handling Hazardous Substances Regulations 1982
NIJB	Northern Ireland Judgements Bulletin (Bluebook)
NLJ	Northern Ireland Legal Journal
NONS	The Notification of New Substances Regulations 1993
npf	Nominal protection factor
NR	Noise rating (curves)
NRA	National Rivers Authority
NRPB	National Radiological Protection Board
NZLR	New Zealand Law Report
OJ	Official journal of the European Community
OECD	Organisation for Economic Development and Co-operation
OES	Occupational exposure standard
OFT	Office of Fair Trading
OR	Operational research

P4SR	Predicted 4 hour sweat rate
Pa	Pascal
PAT	Portable appliance tester
PC	Personal computer
PCB	Polychlorinated biphenyl
PHA	Preliminary hazard analysis
PMNL	Polymorphonuclear leukocyte
PPE	Personal protective equipment
ppm	Parts per million
ptfe	Polytetrafluoroethylene
PTW	Permit to work
PUWER	The Provision and Use of Work Equipment Regulations 1998
PVC	Polyvinyl chloride
QA	Quality assurance
QB	Queen's Bench
QMV	Qualifies majority voting
QUENSH	Quality, environment, safety and health management systems
r.	A clause or regulations of a Regulation
RAD	Reactive airways dysfunction
RCD	Residual cirrent device
RGN	Registered general nurse
RIDDOR	The Reporting of Injuries, Diseases and Dangerous Occurrences Regulations 1995
RM	Resident magistrate
RoSPA	Royal Society for the Prevention of Accidents
RPA	Radiation protection adviser
RPE	Respiratory protective equipment
RPS	Radiation protection supervisor
RR	Risk rating
RRP	Recommended retail price
RSI	Repetitive strain injury
s.	Clause or section of an Act
SAFed	Safety Assessment Federation
SC	Sessions case (in Scotland)
Sen	Sensitizer
SEN	State enrolled nurse
SIESO	Society of Industrial Emergency Services Officers
Sk	Skin (absorption of hazardous substances)
SLT	Scottish Law Times
SMSR	The Supply of Machinery (Safety) Regulations 1992
SPL	Sound pressure level
SRI	Sound reduction index
SRN	State registered nurse
SRSC	The Safety Representatives and Safety Committee Regulations 1977

SSP	Statutory sick pay
Sv	Sievert
SWL	Safe working load
SWORD	Surveillance of work related respiratory diseases
TLV	Threshold Limit Value
TUC	Trades Union Congress
TWA	Time Weighted Average
UEL	Upper explosive limit
UK	United Kingdom
UKAEA	United Kingdom Atomic Energy Authority
UKAS	United Kingdom Accreditation Service
v.	versus
VAT	Value added tax
VCM	Vinyl chloride monomer
vdt	Visual display terminal
VWF	Vibration white finger
WATCH	Working Group on the Assessment of Toxic Chemicals
WBGT	Wet bulb globe temperature
WDA	Waste Disposal Authority
WHSWR	The Workplace (Health, Safety and Welfare) Regulations 1992
WLL	Working load limit
WLR	Weekly Law Report
WRULD	Work related upper limb disorder
ZPP	Zinc protoporphyrin

Appendix 4

Organisations providing safety information

Institution of Occupational Safety and Health, The Grange, Highfield Drive, Wigston, Leicester LE18 1NN 0116 257 3100

National Examination Board in Occupation Safety and Health, NEBOSH, 5 Dominus Way, Meridian Business Park, Leicester LE3 2QW 0116 263 4700 Fax 0116 282 4000

Royal Society for the Prevention of Accidents, Edgbaston Park, 353 Bristol Road, Birmingham B5 7ST 0121 248 2222

British Standards Institution, 389 Chiswick High Road, London W4 4AL 0181 996 9000

Health and Safety Commission, Rose Court, 2 Southwark Bridge, London SE1 9HS 0171 717 6600

Health and Safety Executive, Enquiry Point, Magnum House, Stanley Precinct, Trinity Road, Bootle, Liverpool L20 3QY 0151 951 4000 or any local offices of the HSE

HSE Books, PO Box 1999, Sudbury, Suffolk CO10 6FS 01787 881165

Employment Medical Advisory Service, Daniel House, Trinity Road, Bootle, Liverpool L20 3TW 0151 951 4000

Institution of Fire Engineers, 148 New Walk, Leicester LE1 7QB 0116 255 3654

Medical Commission on Accident Prevention, 35–43 Lincolns Inn Fields, London WC2A 3PN 0171 242 3176

The Asbestos Information Centre Ltd, PO Box 69, Widnes, Cheshire WA8 9GW 0151 420 5866

Chemical Industry Association, King's Building, Smith Square, London SW1P 3JJ 0171 834 3399

Institute of Materials Handling, Cranfield Institute of Technology, Cranfield, Bedford MK43 0AL 01234 750662

National Institute for Occupational Safety and Health, 5600 Fishers Lane, Rockville, Maryland, 20852, USA

Noise Abatement Society, PO Box 518, Eynsford, Dartford, Kent DA4 0LL 01322 862789

Appendices

Home Office, 50 Queen Anne's Gate, London SW1A 9AT 0171 273 4000
Fire Services Inspectorate, Horseferry House, Dean Ryle Street, London SW1P 2AW 0171 217 8728

Department of Trade and Industry: all Departments on 0171 215 5000
Consumer Safety Unit, General Product Safety: 1 Victoria Street, London SW1H 0ET
Gas and Electrical Appliances: 151 Buckingham Palace Road, London SW1W 9SS
Manufacturing Technology Division, 151 Buckingham Palace Road, London SW1W 9SS

Department of Transport
Road and Vehicle Safety Directorate, Great Minster House, 76 Marsham Street, London SW1P 4DR 0171 271 5000

Advisory, Conciliation and Arbitration Service (ACAS), Brandon House, 180 Borough High Street, London SE1 1LW 0171 396 5100

Health Education Authority, Hamilton House, Mabledon Place, London WC1 0171 383 3833

National Radiological Protection Board (NRPB), Harwell, Didcot, Oxfordshire OX11 0RQ 01235 831600

Northern Ireland Office
Health and Safety Inspectorate, 83 Ladas Drive, Belfast BT6 9FJ 01232 701444
Agricultural Inspectorate, Dundonald House, Upper Newtownards Road, Belfast BT4 3SU 01232 65011 ext: 604
Employment Medical Advisory Service, Royston House, 34 Upper Queen Street, Belfast BT1 6FX 01232 233045 ext: 58

Commission of the European Communities, Information Office, 8 Storey's Gate, London SW1P 3AT 0171 222 8122

British Safety Council, National Safety Centre, Chancellor's Road, London W6 9RS 0171 741 1231/2371

Confederation of British Industry, Centre Point, 103 New Oxford Street, London WC1A 1DU 0171 379 7400

Safety Assessment Federation (SAFed), Nutmeg House, 60 Gainsford Street, Butler's Wharf, London SE1 2NY 0171 403 0987

Railway Inspectorate, Rose Court, 2 Southwark Bridge, London SE1 9HS 0171 717 6630

Inspectorate of Pollution, Romney House, 43 Marsham Street, London SW1P 3PY 0171 276 8083

Back Pain and Spinal Injuries Association, Brockley Hill, Stanmore, Middlesex 0181 954 0701

Appendix 5

List of Statutes, Regulations and Orders

Note: This list covers all four volumes of the Series. Entries and page numbers in bold are entries specific to this volume. The prefix number indicates the volume and the suffix number the page in that volume.

Abrasive Wheels Regulations 1970, *1.29*, *1.137*, *4.163*
Activity Centres (Young Persons Safety) Act 1995, *1.58*
Administration of Justice Act 1982, *1.160*
Asbestos (Licensing) Regulations 1983, *4.189*
Asbestos Products (Safety) Regulations 1985, *4.189*
Asbestos (Prohibitions) Regulations 1992, *4.189*

Boiler Explosions Act 1882, *4.130*
Building Act 1984, *1.53*
Building Regulations 1991, *4.56*
Building (Scotland) Act 1959, *4.152*

Carriage of Dangerous Goods by Rail Regulations 1996, *4.217*
Carriage of Dangerous Goods by Road Regulations 1996 (CDG), *4.205*, *4.216–17*
Carriage of Dangerous Goods (Classification, Packaging and Labelling) and Use of Transportable Pressure Receptacles Regulations 1996 (CDG-CPL), *4.204*
Chemical Weapons Act 1996, *4.193*
Chemicals (Hazard Information and Packaging for Supply) Regulations 1994 (CHIP 2), *4.26*, *4.192*, *4.204*, *4.205*
Chemicals (Hazard Information and Packaging for Supply) Regulations 1996 (CHIP 96), *3.91*, *4.192*, *4.204*
Cinematograph (Safety) Regulations 1955, *4.152*
Civil Evidence Act 1968, *1.20*
Civil Liability (Contribution) Act 1978, *1.158*
Classification, Packaging and Labelling of Dangerous Substances Regulations 1984, *4.204*
Clean Air Act 1993, *4.242–3*
Clean Air Acts, *1.56*

Companies Act 1967, *1.53*
Company Directors Disqualification Act 1986, *1.36*
Construction (Design and Management) Regulations 1994, *4.162–3, 4.164*
Construction (Head Protection) Regulations 1989, *4.185–6, 4.201*
Construction (Health, Safety and Welfare) Regulations 1997, *4.165–71, 4.177, 4.179, 4.184–5, 4.186*
Construction (Lifting Operations) Regulations 1961, *4.132, 4.139, 4.141, 4.142–3, 4.144, 4.145, 4.170, 4.178*
Construction (Working Places) Regulations 1966, *1.154*
Consumer Credit Act 1974, *1.120, 1.122*
Consumer Credit (Advertisements) Regulations 1989, *1.121*
Consumer Credit (Total Charge for Credit) Regulations 1980, *1.121*
Consumer Protection Act 1987, *1.39, 1.87, 1.114, 1.118–20, 1.123, 1.124, 1.125, 1.126–7, 1.143, 1.152–3, 2.12, 2.43*
Consumer Transactions (Restrictions on Statements) Order 1976, *1.129*
Contract of Employment Act 1963, *1.82*
Control of Asbestos in the Air Regulations 1990, *4.189*
Control of Asbestos at Work (Amendment) Regulations 1992, *4.189*
Control of Asbestos at Work Regulations 1987, *4.150, 4.201*
Control of Industrial Major Accident Hazards Regulations 1984 (CIMAH), *4.202, 4.203–4, 4.231*
Control of Lead at Work Regulations 1980, *1.82, 4.151, 4.201*
Control of Major Accident Hazards and Regulations (COMAH) (proposed), *4.203, 4.212*
Control of Misleading Advertisements Regulations 1988, *1.114, 1.127, 1.128*
Control of Pollution Act 1974, *1.46, 1.142, 2.12, 3.135, 4.239*
Control of Pollution (Special Waste) Regulations 1980, *4.244*
Control of Substances Hazardous to Health Regulations 1994 (COSHH), *1.41, 1.82, 2.4, 2.5, 2.12, 3.8, 3.87, 3.88–9, 3.91–2, 3.155–6, 3.158, 4.132, 4.150, 4.151, 4.192, 4.195–7, 4.201, 4.205, 4.206–8, 4.228*
Controlled Waste Regulations 1992, *4.240, 4.242*
Courts and Legal Services Act 1990, *1.23*
Criminal Justice and Public Order Act 1994, *1.21*
Criminal Justices Act 1991, *1.52*

Dangerous Substances (Notification and Marking of Sites) Regulations 1990, *4.231*
Disability Discrimination Act 1995, *1.90, 1.95–6, 3.190, 3.193*
Disability Discrimination (Employment) Regulations 1996, *1.96*
Diving at Work Regulations 1997, *4.176*
Docks Regulations 1934, *4.141*
Docks Regulations 1988, *4.132, 4.141, 4.142–3, 4.145, 4.146*

EEC/EU Directives
  banning the use of methyl chloroform, *3.51*
  exposure to noise at work, *3.136*

general product safety directive, *1.123, 1.124*
lifts directive, *4.97*
machinery directive, *1.66, 4.69*
minimum safety and health requirements for the workplace, *4.63*
product liability, *1.143*
protection of young people at work, *1.97*
provision and control of shipments of waste within, into and out of the European Community, *4.238*
safety and health of workers, *1.36, 4.63*
work equipment directive, *1.70, 4.139*
workplace directive, *1.70, 1.138*
Electricity at Work Regulations 1989, *4.75, 4.110–1, 4.116, 4.118–22, 4.123–4, 4.127, 4.128, 4.132, 4.152, 4.171, 4.172*
Electricity at Work Regulations (Northern Ireland) 1991, *4.110*
Employer's Liability Act 1880, *1.133*
Employer's Liability (Compulsory Insurance) Act 1969, *1.76, 1.136*
Employer's Liability (Compulsory Insurance) General Regulations 1972, *1.136*
Employer's Liability (Defective Equipment) Act 1969, *1.147, 1.155, 1.156*
Employment Act 1989, *1.83*
Employment of Children Act 1973, *1.78*
Employment Protection (Consolidation) Act 1978, *1.90*
Employment Rights Act 1996 (ERA), *1.90, 1.91–2, 1.102–3, 1.107–8, 1.109–10, 1.111*
Environment Act 1995, *1.56, 4.209–11*
Environmental Protection Act 1990, *1.56, 1.142, 4.208–12, 4.239–41, 4.243, 4.246*
Environmental Protection (Duty of Care) Regulations 1991, *4.241*
Environmental Protection (Prescribed Processes and Substances) Regulations 1991, *4.209, 4.240*
Equal Pay Act 1970, *1.30, 1.83, 1.90, 1.96*
Equal Pay (Amendment) Regulations 1983, *1.83*
European Communities Act 1972, *1.31*
European Communities Act 1992, *1.31*
European Community/Union Directives, *1.36, 1.41, 1.47, 4.68*
European Council
  Framework Directive 1989, *1.98*
  Safety Standards Against the Dangers of Ionising Radiation Directive, *3.114*
  Working Time Directive 1993, *1.99*
Explosives Act 1875, *1.56, 2.68*

Factories Act 1833, *3.4*
Factories Act 1846, *1.133*
Factories Act 1937, *1.34*
Factories Act 1961, *1.34, 1.35, 1.37, 1.38, 1.54, 1.137, 1.138, **1.154, 1.155**, 2.17, 2.43, 4.68, 4.131, 4.132, 4.134, 4.139, 4.140, 4.141, 4.142, 4.143, 4.145, 4.154–5, 4.185, 4.186*
Factories and Workshops Act 1901, *4.131*

Fatal Accidents Act 1976, *1.146*, *1.160*
Fire Certificates (Special Premises) Regulations 1996, *4.63*, *4.186–7*
Fire Precautions Act 1971, *1.53*, *1.54–6*, *2.11*, *4.60–3*
Fire Precautions (Workplace) Regulations 1997, *1.50*, *1.55*, *2.11*, *3.190*, *4.63*
Fire Safety and Safety of Places of Sport Act 1987, *1.55*, *4.60*
Fire Services Act 1971, *4.60*
Food and Drugs Act 1955, *1.39*
Food Safety Act 1990, *1.39*, *1.58*, *1.126*
Food Safety (General Food Hygiene) Regulations 1995, *1.58*, *4.188*

General Product Safety Regulations 1994, *1.114*, *1.122*, *1.123*
Goods Vehicles (Plating and Testing) Regulations 1988, *1.57*

Health and Safety at Work etc. Act 1974 (Application outside Great Britain) Order 1995, *1.54*
Health and Safety at Work etc. Act 1974 (HSW)
  Parts *II–IV*, *1.53*
  s.2(2)(e), *1.40*
  s.2–6, *1.35*, *1.52*
  s.2–9, *1.49*
  s.2, *1.5*, *1.46*, *1.47*, *2.158*
  s.3, *1.47*
  s.5, *1.47*, *1.50*
  s.6, *1.87*, *2.41*, *2.43*
  s.7–8, *1.48*, *1.52*
  3s.9, *1.48*
  s.10, *1.48*
  s.11–14, *1.49*
  s.11, *1.48*
  s.15, *1.35*, *1.49*, *1.63*
  s.16(2), *1.65*
  s.16, *1.29*, *1.49*
  s.17(2), *1.50*
  s.17, *1.29*
  s.18, *1.48*, *1.50*
  s.19, *1.50*
  s.20, *1.50*
  s.21–25, *1.58*
  s.21, *1.51*
  s.22, *1.51*
  s.25A, *1.51*
  s.33, *1.36*, *1.51–2*
  s.34, *1.52*
  s.37, *1.4*, *1.36*, *1.52*
  s.38, *1.4*
  s.39, *1.4*
  s.40, *1.5*, *1.52*
  s.48, *1.58*

s.50, *1.29*, *1.31*, *1.64*
s.53, *1.47*
s.80, *1.29*
s.82, *1.29*
schedule 1, *1.35*, *1.45*
schedule 3, *1.29*, *1.35*, *1.49*
schedules, *1.53*
Section 1(2), *1.36*
Section 2–6, *1.36*
Section 6, *2.12*
Section 9, *1.37*
Section 15, *1.29*, *1.63–4*
Section 38, *1.18*
Section 47, *1.39*
Health and Safety at Work etc. Act 1974 (HSW), *1.3*, *1.7*, *1.8*, *1.11*, *1.13*, *1.20*, *1.25*, *1.29*, *1.34*, *1.35–6*, *1.37*, *1.41*, *1.44–54*, *1.62*, *1.64*, *1.82*, *1.86*, *1.142*, *1.156*, *2.3*, *2.17*, *2.43*, *2.143*, *2.156*, *2.158*, *3.9*, *3.114*, *4.108–10*, *4.131*, *4.161*, *4.163*, *4.164*, *4.166*
Health and Safety at Work (NI) Order 1978, *1.30*
Health and Safety (Consultation with Employees) Regulations 1996, *1.98*, *2.3*
Health and Safety (Display Screens Equipment) Regulations 1992 (DSER), *1.49*, *2.158*, *3.170*, *3.192*
Health and Safety (First Aid) Regulations 1981, *3.9*
Health and Safety (Leasing Arrangements) Regulations 1980, *1.87*
Health and Safety (Safety Signs and Signals) Regulations 1996, *4.58*
Health and Safety (Young Persons) Regulations 1997, *1.96–7*
Highly Flammable Liquids and Liquefied Petroleum Gases Regulations 1972, *2.11*, *4.26*
Hoists Exemption Order 1962, *4.98*, *4.140*

Interpretation Act 1978, *1.30*
Ionising Radiations Regulations 1985, *1.82*, *3.114–7*, *4.201*

Justices of the Peace Act 1979, *1.23*

Law Reform (Contributory Negligence) Act 1945, *1.146*, *1.157*
Law Reform (Miscellaneous Provisions) Act 1934, *1.160*
Lifting Operations and Lifting Equipment Regulations 1998 (LOLER), *1.54*, *1.66*, *4.95*, *4.97–9*, *4.99*, *4.132*, *4.139*, *4.140*, *4.147–8*, *4.178*
Lifting Plant and Equipment (Record of Test and Examination etc.) Regulations 1992, *4.139*, *4.140*, *4.146–7*
Lifts Regulations 1997 (LR), *4.70*, *4.95*, *4.96–7*, *4.132*, *4.140*, *4.141*
Limitation Act 1980, *1.158–9*
Limitation Acts 1939, 1975, 1980, *1.139*
Loading and Unloading of Fishing Vessels Regulations 1988, *4.139*, *4.142*, *4.145*

Management and Administration of Safety and Health in Mines Regulations 1993, *1.56*
Management of Health and Safety at Work Regulations 1992 (MHSW), *1.36*, *1.39*, *1.49*, *1.96*, *1.138*, 2.3, 2.5, 2.6, 2.22, 2.43, 3.193, 4.163, 4.164, 4.185, 4.194, 4.230–1
Manual Handling Operations Regulations 1992 (MHOR), *1.49*, 2.158, 2.161, 2.166, 3.192, 4.198
Medicine Act 1968, *1.126*
Merchant Shipping Act 1988, *1.26*
Mines Management Act 1971, *1.56*
Mines and Quarries Act 1954, *1.35*, *1.56*, 4.188–9
Mines and Quarries (Tips) Act 1969, *1.56*
Miscellaneous Mines Order 1956, 4.139, 4.141, 4.142
Misrepresentation Act 1967, *1.79*, *1.114*
Motor Vehicle (Construction and Use) Regulations 1986, *1.57*
Motor Vehicle (Tests) Regulations 1982, *1.57*
Motor Vehicles (Compulsory Insurance) Regulations 1992, *1.137*

National Health Service (Amendment) Act 1986, *1.58*
National Insurance (Industrial Injuries) Act 1946, *1.133*
Noise at Work Regulations 1989, *1.141*, 3.96, 3.134, 3.136–7, 4.185, 4.201
Noise and Statutory Nuisance Act 1993, 4.243
Northern Ireland (Emergency Provisions) Acts 1978 and 1987, *1.4*
Notification of Installations Handling Hazardous Substances Regulations 1982(NIIHHS), 4.202–3
Notification of New Substances Regulations 1993, 4.193

Occupier's Liability Act 1957, *1.40*, *1.86*, *1.142*, *1.147*, *1.151–2*
Occupier's Liability Act 1984, *1.40*, *1.86*, *1.86–7*, *1.142*, *1.147*, *1.151–2*
Offices Act 1960, *1.35*
Offices, Shops and Railway Premises Act 1963, *1.35*, 4.186
Offices, Shops and Railway Premises (Hoists and Lifts) Regulations 1968, 4.132, 4.139, 4.140
Offshore (Electricity and Noise) Regulations 1997, 4.110
Offshore Safety Act 1992, *1.52*

Packaging, Labelling and Carriage of Radioactive Material by Rail Regulations 1996, 4.217
Personal Protective Equipment at Work Regulations 1992, *1.49*, 3.97, 3.192, 4.117, 4.185, 4.199
Petroleum (Consolidation) Act 1928, *1.56*, *1.57*, *1.58*, 4.152, 4.188
Petroleum Mixtures Order 1929, 4.188
Petroleum Spirit (Motor Vehicles etc.) Regulations 1929, 4.188
Planning (Hazardous Substances) Regulations 1992, 4.231
Power Press Regulations 1965, 4.132, 4.148–9
Powers of Criminal Courts Act 1973, *1.118*, *1.120*
Pressure Systems and Transportable Gas Container Regulations 1989, *1.66*, 4.100, 4.102–3, 4.131, 4.132, 4.133–4, 4.171
Price Indications (Bureaux de Change) (No.2) Regulations 1992, *1.120*

Price Indications (Method of Payment) Regulations 1990, *1.120*
Price Marking Order 1991, *1.120*
Protection of Eyes Regulations 1974, *4.185*
Provision and Use of Work Equipment Regulations 1992 (PUWER), *1.37*, *1.38*, *1.49*, *1.70*, *1.138*, *3.191*, *4.163*, *4.171*
Provision and Use of Work Equipment Regulations 1998 (PUWER 2), *4.69*, *4.73–6*, *4.90*, *4.91–3*, *4.95*, *4.96–9*
Public Health Act 1936, *4.44*

Quarries (General) Regulations 1956, *4.132*
Quarries Order 1956, *4.139*, *4.141*, *4.142*

Race Relations Act 1976, *1.83*, *1.90*, *1.94–5*
Radioactive Material (Road Transport) Act 1991, *3.118*
Radioactive Substances Act 1960, *1.56*
Radioactive Substances Act 1993, *3.117*
Radiological Protection Act 1970, *1.53*
Redundancy Payments Act 1965, *1.159*
Rehabilitation of Offenders Act 1974, *1.106*
Reporting of Injuries, Diseases and Dangerous Occurrences Regulations 1995 (RIDDOR), *2.65*, *2.67*, *3.69*, *3.193*
Road Traffic Act 1972, *1.142*, *2.68*
Road Traffic Acts 1972–91, *1.57*
Road Traffic Regulation Act 1984, *1.57*
Road Vehicle Lighting Regulations 1981, *1.57*

Safety Representatives and Safety Committee Regulations 1977 (SRSC), *1.98*, *2.3*
Sale of Goods Act 1893, *1.85*, *1.143*
Sale of Goods Act 1979, *1.79*, *1.84*, *1.121–2*, *1.143*
Sale of Goods Act 1994, *1.121*
Sex Discrimination Act 1975, *1.83*, *1.90*, *1.92–4*, *1.94*
Sex Discrimination Act 1986, *1.94*
Ship Repairing Regulations 1988, *4.141*
Shipbuilding and Ship Repairing Regulations 1960, *4.132*, *4.139*, *4.143*, *4.145*
Single European Act 1985, *1.31*
Single European Act 1986, *1.27*, *1.33*, *1.36*, *1.67*
Social Security Act 1989, *1.135*, *1.145*, *1.159*
Social Security (Claims and Payments) Regulations 1979, *2.69*
Social Security and Housing Benefit Act 1982, *1.134*
Social Security (Industrial Injuries and Diseases) Miscellaneous Provisions Regulations 1986, *1.134*
Social Security (Industrial Injuries) (Prescribed Diseases) Amendment Regulations 1985, *1.141*
Social Security (Industrial Injuries) (Prescribed Diseases) Regulations 1975, *1.140*
Social Security (Industrial Injuries) (Prescribed Diseases) Regulations 1985, *1.135*, *1.140*
Social Security (Recoupment) Regulations 1990, *1.145*

Social Security (Recovery of Benefits) Act 1997, *1.145*
Special Waste Regulations 1996, *4.211*, *4.240*, *4.244–5*
Supply of Goods (Implied Terms) Act 1973, *1.122*, *1.143*
Supply of Goods and Services Act 1982, *1.85*, *1.122*
Supply of Machinery (Safety) Regulations 1992 (SMSR), *1.70*, *3.192*, *4.69–73*, *4.95*, *4.96*, *4.99*, *4.139*, *4.141*, *4.143*, *4.144*, *4.145*

Trade Descriptions Act 1968, *1.114*, *1.115–18*, *1.119*
Trade Union and Employment Rights Act 1993, *1.90*
Transfer of Undertakings (Protection of Employment) Regulations 1981, *1.90*
Transfrontier Shipment of Hazardous Waste Regulations 1994, *4.238*
Treaty of European Union 1991, *1.31*
Treaty of Rome 1957, *1.31*

Unfair Contract Terms Act 1977, *1.85*, *1.86*, *1.128*, *1.129*, *1.143*, *1.151*, *2.8*
Unfair Terms in Consumer Contracts Regulations 1994, *1.85*, *1.114*, *1.128*, *1.129–30*
Units of Measurement Regulations 1980, *3.108*

Wages Act 1986, *1.90*
Waste Management Licensing (Amendments etc.) Regulations 1995, *4.244*
Waste Management Licensing Regulations 1994, *4.243*
Water Act 1989, *1.56*
Water Industry Act 1991, *4.242*
Water Resources Act 1991, *4.241–2*
Work in Compressed Air Special Regulations 1958, *4.176*
Workmen's Compensation Act 1897, *1.133*, *3.5*
Workmen's Compensation Act 1925, *1.133*
Workplace (Health Safety and Welfare) Regulations 1992 (WHSWR), *1.7*, *1.39*, *1.49*, *1.138*, *3.191*

Appendix 6

List of Cases

Note: This list covers all four volumes of the Series. Entries and page numbers in bold are entries specific to this volume. The prefix number indicates the volume and the suffix number the page in that volume.

A.G. Stanley Ltd v. Surrey County Council (1994) 159 JP 691, *1.119*
AG v. Times Newspapers Ltd (1979) 2 EHRR 245, European Court of Human Rights, *1.28*
Allen v. Redbridge (1994) 1 All ER 728, *1.120*
Ashdown v. Samuel Williams & Sons (1957) 1 All ER 35, *1.86*
Ashington Piggeries Ltd v. Christopher Hill Ltd (1971) 1 All ER 847, *1.84*
Ashley v. Sutton London Borough Council (1994) 159 JP 631, *1.117*
Austin v. British Aircraft Corporation Ltd (1978) IRLR 332, *1.106*

Baker v. T.E. Hopkins & Sons Ltd (1959) 1 WLR 966, *1.158*
Balfour v. Balfour (1919) 2 KB 571, *1.78*
Ball v. Insurance Officer (1985) 1 All ER 833, *1.134*
Beale v. Taylor (1967) 3 All ER 253, *1.84*
Beckett v. Cohen (1973) 1 All ER 120, *1.117*
Berry v. Stone Manganese Marine Ltd (1972) 1 Lloyd's Reports 182, *1.140*
Bett v. Dalmey Oil Co. (1905) 7F (Ct of Sess.) 787, *2.157*
Bigg v. Boyd Gibbins Ltd (1971) 2 All ER 183, *1.78*
Bowater v. Rowley Regis Corporation (1944) 1 All ER 465, *1.139*
British Airways Board v. Taylor (1976) 1 All ER 65, *1.117*
British Coal v. Armstrong and Others (*The Times*, 6 December 1996, CA), *1.141*
British Gas Board v. Lubbock (1974) 1 WLR 37, *1.115*
British Home Stores v. Burchell (1978) IRLR 379, *1.106*
British Railways Board v. Herrington (1971) 1 All ER 897, *1.142*
Buck v. English Electric Co. Ltd (1978) 1 All ER 271, *1.139*
Bulmer v. Bollinger (1974) 4 All ER 1226, *1.26*
Bunker v. Charles Brand & Son Ltd (1969) 2 All ER 59, *1.151*

Cadbury Ltd v. Halliday (1975) 2 All ER 226, *1.115*
Carlill v. Carbolic Smoke Ball Co. (1893) 1 QB 256, *1.78*
Close v. Steel Company of Wales (1962) AC 367, *1.38*
Cunningham v. Reading Football Club (1991) *The Independent*, 20 March 1991, *1.40*

Darbishire v. Warren (1963) 3 All ER 310, *1.80*
Davie v. New Merton Board Mills Ltd (1959) 1 All ER 67, *1.156*
Davies v. De Havilland Aircraft Company Ltd (1950) 2 All ER 582, *1.138*
Director General of Fair Trading v. Tyler Barrett and Co. Ltd (1 July 1997, unreported), *1.127*
Dixons Ltd v. Barnett (1988) BTLC 311, *1.116*
Donoghue (McAlister) v. Stevenson (1932) All ER Reprints 1, *1.143*
Donoghue v. Stevenson (1932) AC 562, 38, 149, *1.152*
Dunlop Pneumatic Tyre Co. Ltd v. Selfridge & Co. Ltd (1915) AC 847, *1.78*

East Lindsay District Council v. Daubny (1977) IRLR 181, *1.104*
Edwards v. National Coal Board (1949) 1 KB 704, (1949) 1 All ER 743, *2.157, 4.198*
European Court of Justice cases
   C382/92 Safeguarding of employee rights in the event of transfer of undertakings, Celex no. 692JO382, EU Luxembourg (1992), *1.98*
   C383/92 Collective Redundancies, Celex no. 692JO383, EU Luxembourg (1992), *1.98*

Factortame Ltd No. 5, Times Law Reports, 28 April 1998, *1.26*
Fenton v. Thorley & Co. Ltd (1903) AC 443, *2.4*
Fitch v. Dewes (1921) 2 AC 158, *1.80*
Fitzgerald v. Lane (1988) 3 WLR 356, *1.151, 1.157*
Fletcher Construction Co. Ltd v. Webster (1948) NZLR 514, *1.150*
Frost v. John Summers and Son Ltd (1955) 1 All ER 870, *1.137*

General Cleaning Contractors Ltd v. Christmas (1952) 2 All ER 1110, *1.138*
General Cleaning Contractors Ltd v. Christmas (1953) AC 180, *1.154*
George Mitchell (Chesterhall) Ltd v. Finney Lock Seeds Ltd (1983) 2 All ER 737, *1.85*
Global Marketing Europe (UK) Ltd v. Berkshire County Council Department of Trading Standards (1995) Crim LR 431, *1.117*

Hadley v. Baxendale (1854) 9 Exch. 341, *1.80*
Heal v. Garringtons, unreported, 26 May 1982, *1.141*
Hedley Byrne & Co. Ltd v. Heller & Partners Ltd (1964) AC 463, *1.17, 1.38*
Henderson v. Henry E. Jenkins & Sons (1969) 3 All ER 756, *1.138*
Henry Kendall & Sons v. William Lillico & Sons Ltd (1968) 2 All ER 444, *1.84*

Hicks v. Sullam (1983) MR 122, *1.116*
Hivac Ltd v. Park Royal Scientific Instruments Ltd (1946) 1 All ER 350, *1.83*
Home Office v. Holmes (1984) IRLR; (1984) 3 All ER 549; (1985) 1 WLR 71, *1.93*
Hutchinson v. Enfield Rolling Mills (1981) IRLR 318, *1.104*

International Sports Co. Ltd v. Thomson (1980) IRLR 340, *1.104*

James v. Eastleigh Borough Council (1990) IRLR 288, *1.92*
James v. Waltham Cross UDC (1973(IRLR 202, *1.105*
John Summers & Sons Ltd v. Frost (1955) AC 740, *1.29*
Joseph v. Ministry of Defence Court of Appeal Judgement 29 February 1980 – *The Times* 4 March 1980, *1.141*
Junior Books Co. Ltd v. Veitchi (1983) AC 520, *1.152*

Keys v. Shoefayre Ltd (1978) IRLR 476, *1.107*
Kilgullan v. W. Cooke and Co. Ltd (1956) 2 All ER 294, *1.138*

Latimer v. AEC Ltd (1953) 2 All ER 449, *1.149*
Lim Poh Choo v. Camden and Islington Area Health Authority (1979) 2 All ER 910, *1.145*
Lindsay v. Dunlop Ltd (1979) IRLR 93, *1.107*
Lister v. Romford Ice and Cold Storage Co. Ltd (1957) AC 535, *1.83*

McDermid v. Nash Dredging and Reclamation Co.Ltd (1987) 3 WLR 212, *1.153*
McGuiness v. Kirkstall Forge Engineering Ltd QBD Liverpool 22 February 1979 (unreported), *1.140*
McKew v. Holland and Hannan and Cubitts Ltd (1969) 2 All ER 1621, *1.151*
McNeil v. Dickson and Mann (1957) SC 345, *4.155*
Malik v. BCCI (1997) 3 All ER 1, *1.84*
Marshall v. Southampton and South West Hampshire Area Health Authority (Teaching) (1986) case 152/84 1 CMLR 688; (1986) QB 401, *1.34*
Matthews v. Kuwait Bechtal Corporation (1959) 2 QB 57, *1.79*
MFI Furniture Centres Ltd v. Hibbert (1995) 160 JP 178, *1.118*
MFI Warehouses Ltd v. Nattrass (1973) 1 All ER 762, *1.117*
Moeliker v. Reyrolle WLR 4 February 1977, *1.145*
Mullard v. Ben Line Steamers Ltd (1971) 2 All ER 424, *1.146*

Nancollas v. Insurance Officer (1985) 1 All ER 833, *1.134*
Newham London Borough v. Singh (1987) 152 JP, *2.77*
Nordenfelt v. Maxim Nordenfelt Guns and Ammunition Co. (1894) AC 535, *1.80*
Norman v. Bennett (1974) 3 All ER 351, *1.118*

Ollett v. Bristol Aerojet Ltd (1979) 3 All ER 544, *1.144*

Page v. Freight Hire Tank Haulage Ltd (1980) ICR 29; (1981) IRLR 13, *1.94*
Paris v. Stepney Borough Council (1951) AC 367, *1.149*
Parsons v. B.N.M. Laboratories Ltd (1963) 2 All ER 658, *1.80*
Pickstone v. Freeman plc (1989) 1 AC 66, *1.30*
Pitts v. Hill and Another (1990) 3 All ER 344, *1.146*
Planche v. Colburn (1831) 8 Bing 14, *1.80*
Polkey v. A.E. Dayton (Services) Ltd (1988) IRLR 503; (1987) All ER 974, HE (E), *1.101, 1.108*

Queensway Discount Warehouses v. Burke (1985) BTLC 43, *1.115*
Quintas v. National Smelting Co. Ltd (1961) 1 All ER 630, *1.138*

R. v. Bevelectric (1992) 157 JP 323, *1.117*
R. v. British Steel plc (1995) ICR 587, *1.37*
R. v. Bull, *The Times*, 4 December 1993, *1.118*
R. v. George Maxwell Ltd (1980) 2 All ER 99, *1.8*
R. v. Kent County Council (6 May 1993, unreported), *1.118*
R. v. Secretary of State for Transport v. Factortame Ltd C 221/89; (1991) 1 AC 603; (1992) QB 680, *1.26, 1.27*
R. v. Sunair Holidays Ltd (1973) 2 All ER 1233, *1.117*
R. v. Swan Hunter Shipbuilders Ltd and Telemeter Installations Ltd (1981) IRLR 403, *4.163–8*
Rafdiq Mughal v. Reuters (1993), *3.185–4*
Readmans Ltd v. Leeds City Council (1992) COD 419, *1.25*
Ready-mixed Concrete (South East) Ltd v. Minister of Pensions and National Insurance (1968) 1 All ER 433, *1.81*
Roberts v. Leonard (1995) 159 JP 711, *1.117*
Rowland v. Divall (1923) 2 KB 500, *1.84*
R.S. Components Ltd v. Irwin (1973) IRLR 239, *1.109*
Rylands v. Fletcher (1861) 73 All ER Reprints N. 1, *1.142*

Sanders v. Scottish National Camps Association (1980) IRLR 174, *1.109*
Scammell v. Ouston (1941) All ER 14, *1.78*
SCM (UK) Ltd v. W. J. Whittle and Son Ltd (1970) 2 All ER 417, *1.145*
Scott v. London Dock Company (1865) 3 H and C 596, *1.139*
Shepherd v. Firth Brown (1985) unreported, *1.141*
Sillifant v. Powell Duffryn Timber Ltd (1983) IRLR 91, *1.101*
Smith v. Baker (1891) AC 325, *1.158*
Smith v. Crossley Bros. Ltd (1951) 95 Sol. Jo. 655, *1.153*
Smith v. Leach Brain & Co. Ltd (1962) 2 WLR 148, *1.150*
Smith v. Stages (1989) 1 All ER 833, *1.137*
Spartan Steel and Alloys Ltd v. Martin and Co. (Contractors) Ltd (1972) 3 All ER 557, *1.145*
Spencer v. Paragon Wallpapers Ltd (1976) IRLR 373, *1.103*
Stevenson, Jordan and Harrison v. Macdonald & Evans (1951) 68 R.P.C. 190, *1.81*
Systems Floors (UK) Ltd v. Daniel (1982) ICR 54; (1981) IRLR 475, *1.82*

Taylor v. Alidair Ltd (1978) IRLR 82, 104–5
Tesco Supermarkets v. Nattrass (1972) AC 153, *1.116*
Thompson v. Smiths Ship Repairers (North Shields) Ltd (1984) 1 All ER 881, *1.140*
Toys R Us v. Gloucestershire County Council (1994) 158 JP 338, *1.118*
Treganowan v. Robert Knee & Co. Ltd (1975) IRLR 247; (1975) ICR 405, *1.109*

Vandyke v. Fender (1970) 2 All ER 335, *1.137*
Victoria Laundry (Windsor) Ltd v. Newman Industries Ltd (1949) 2 KB 528, *1.80*

Walton v. British Leyland (UK) Ltd (12 July 1978, unreported), *1.124*
Ward v. Tesco Stores (1976) 1 All ER 219, *1.139*
Waugh v. British Railways Board (1979) 2 All ER 1169, 20, *1.144*
Wheat v. Lacon & Co. Ltd (1966) 1 All ER 35, *1.87*
Williams v. Compair Maxam Ltd (1982) ICR 800, *1.108*
Wilsher v. Essex Health Authority (1989) 2 WLR 557, *1.151*
Wilson v. Rickett, Cockerell & Co. Ltd (1954) 1 All ER 868, *1.84*
Wilsons and Clyde Coal Co. Ltd v. English (1938) AC 57 (HL), *2.18*, *2.157*
Wing Ltd v. Ellis (1985) AC 272, *1.118*

Young v. Bristol Aeroplane Company Ltd (1994) 2 All ER 293, *1.133*

Appendix 7

Series Index

Note: This list covers all four volumes of the Series. Entries and page numbers in bold are entries specific to this volume. The prefix number indicates the volume and the suffix number the page in that volume.

Abatement notices, chemicals nuisances, 4.212
ACAS (Advisory, Conciliation and Arbitration Service), advisory handbook, 1.99–1.101
Accident control systems, 2.44
Accident costing/cost management, 2.47–2.49, 2.52–2.53
Accident investigations:
  information collection:
    interviewing for, 2.81–2.84
    what happened, 2.80–2.84
    why it happened, 2.84–2.88
  purpose, 2.79–2.80
  report writing, 2.88–2.90
Accident prevention:
  Accident Prevention Advisory Unit (APAU), 2.51
  economic reasons for, 2.18–2.19
  humanitarian reasons, 2.17–2.18
  legal reasons, 2.17
Accident/incident data:
  accident statistics, 2.23
  accident triangles, 2.59–2.61
  checking non-reporting, 2.63–2.64
  collection of, 2.61–2.65
  ensuring reporting of, 2.62–2.63
  epidemiological analysis, 2.78–2.79
  frequency/severity relationship, 2.59–2.60
  recording details, 2.64
  for risk measurement, 2.77
  trend analysis, 2.69–2.77
  types of, 2.57–2.61
  use of, 2.69–2.77
Accidents:
  causation models, 2.14–2.16
  definition, 2.3–2.4, 2.57
  electrical, 4.111–4.112
  failure modes, 2.16
  fatal, 1.160–1.161
  fault tree analysis, 2.16
  and incidents, 2.3–2.4
  inevitable, 1.139
  modelling, 2.36
  multiple causation theory, 2.15–2.16
  prevention of, 2.3–2.4
  sex/ethnic differences in liability, 2.131
  susceptibility with change, 2.131–2.132
  UK notification requirements, 2.65–2.69
  use of data on, 2.44–2.45
  at work, benefits available, 1.134–1.136
  zero tolerance to, 2.143–2.144
  see also Incident
Accommodation, of eye, 3.164
Acid rain, 4.236
Acids and bases, 4.7–4.8
Acne, 3.34
Acquired immune deficiency syndrome (AIDS), 3.66
Activity centres, young persons, 1.58
Acts of Parliament, 1.29–1.30
Actus reus, **1.37**
Administrative law, 1.5–1.7
Adopted legislation, 1.31–1.33
Adrenaline, 3.23
Adversary law system, 1.18
Advertising:
  Advertising Standards Authority, 1.127
  broadcast, 1.128
  misleading, 1.127–1.128
  truth in, 1.120–1.121
Advisory Committee of Toxic Substances (ACTS), 3.88
Advisory services, health, 3.6
Affidavits, 1.22
AIDS (acquired immune deficiency syndrome), 3.66
Air, composition, 4.9
Air pollution/sampling/control:
  air pollution control (APC), 4.240
  atomic absorption spectrometry, 3.81
  breathing apparatus, 3.101
  control measures, 3.94–3.103
  direct monitoring, 3.82–3.84
  dust, 3.74, 3.81, 3.82, 3.83–3.84, 3.96, 3.150–3.152
  environmental legislation, 4.242–4.243

fibre monitoring, 3.81–3.82
fibres, 3.81, 3.83–3.84
gas chromatography, 3.80
gases and vapours, measurement, 3.75–3.78, 3.82–3.83, 3.153
grab, 3.75–3.78
hygrometry, 3.85–3.86
Inspectorate of Pollution, 1.56
Integrated Pollution Control (IPC), 1.56, 4.240
international standards, 3.93
limit changes, 3.93–3.94
long-term, 3.78–3.82
Maximum Exposure Limits (MELs), 3.87–3.88
mixtures, 3.90
neurophysiological changes, 3.93
Occupational Exposure Standards (OESs), 3.87–3.88, 3.89
oxygen analysers, 3.84–3.85
personal protective equipment (PPE), 3.97–3.98
planned strategies, 3.78
reduced time exposure, 3.96
respirators, 3.99
safety specification, 3.94
segregation, 3.95
stain detector tubes, 3.75–3.80
Threshold Limit Values (TLV), 3.87, 3.92–3.93
Time Weighted Average concentrations (TWA), 3.89
time weighted average (TWA), 3.89, 3.150, 3.155
toxic substance substitution, 3.94–3.95
UK exposure limits, 3.87–3.90
ventilation:
  dilution, 3.96, 3.149
  local extract, 3.95, 3.149
Air receivers *see* Steam and air receivers
Air velocity measurement, 3.158–3.161
Alkyl mercury, disease from, 3.44
All England Reports (All ER), 1.18
Alpha radiation, 3.106
Alternative dispute resolution, 1.22
Alveolitis, extrinsic allergic, 3.41–3.42
American Conference of Governmental and Industrial Hygienists (ACGIH), 3.87, 3.155
American Industrial Hygiene Association, 3.74
Ammonia, effects of, 3.53
Anatomy *see* Human body, structure and functions
Angiosarcoma, 3.57
Animal handler's lung, 3.42
Antecedent/behaviour/consequence (ABC) model, 2.146–2.148
Anthrax, 3.67–3.68
Anthropometry, 3.183–3.184
Aplasia, 3.25

Appeal, right of, 1.101
Appellate Committee, 1.23
Appellate jurisdiction, 1.7
Approved codes, status of, 1.50
Approved Codes of Practice (ACOP), 1.67
  ventilation, 3.156
Aptitude tests, 2.107
Arc welding, fumes from, 3.95
Arsenic, skin cancer from, 3.45
Arsine, diseases from, 3.45
Arthritis, 3.190
Artificial respiration, 3.16
Asbestos:
  construction sites, 4.189
  disposal legislation, 4.189, 4.245
Asbestos dust control, and local exhaust ventilation (LEV), 4.150–4.151
Asbestosis, 3.27–3.28, 3.39
Asphyxia, simple and toxic, 3.52
Association of British Insurers, 1.85
Astigmatism, 3.164
Atkinson, J.W., 2.101
Atomic absorption spectrometry, 3.81
Atomic numbers, 4.4
Attendance Allowance, 1.135
Attitudes, 2.129–2.130
  and behaviour, 2.143–2.144
Attorney General, 1.4, 1.23, 1.24
Audits, independent, 2.22–2.23

Back problems, 3.188–3.190
Bagassosis, 3.42
Barometer, mercury, 4.10
Barrier creams, 3.34, 3.102–3.103
Barristers, and representation, 1.8
Bases and acids, 4.7–4.8
Basle Convention, 4.238–4.239
Behaviour *see* Human function and behaviour
Behavioural science:
  aims of, 2.96
  and health and safety, 2.97–2.98
Bends, 3.16
Benzene, toxic effects, 3.51
Beryllium dust, Pneumoconiosis from, 3.37
Best available techniques not entailing excessive costs (BATNEEC), 4.240–4.241
Best practical environmental option (BPEO), 4.241
Beta radiation, 3.106–3.107
Bills, House of Commons, 1.29–1.30
Binet, A., 2.107
Biological agents, as hazards, 3.91–3.92
Biological danger, immediate/long-term, 2.14
Bird fancier's lung, 3.42
Bird, F.E., 2.9, 2.10, 2.15, 2.32, 2.38, 2.44, 2.52, 2.55
Bladder, and cancer, 3.71

Blindness, temporary, 3.164
Blood cells, red and white, 3.18
Boilers:
  competent persons, 4.155
  defects, typical, 4.136–4.137
  definition, 4.134
  examinations, 4.135–4.136, 4.155
  patent and latent defects, 4.154–4.155
  reports, 4.155–4.156
  safe operating limits, 4.134
  safety valves, 4.134
  sound material, 4.155
Bone, osteolysis and necrosis, 3.71
Brain, 3.20
  brain damage, 3.16
  narcosis and encephalopathy, 3.71
Brainstorming, 2.136
Breach of statutory duty, 1.38, 1.39
Breathing apparatus, types of, 3.100, 3.101
Bremsstrahlung radiation, 3.106, 3.107
British Occupational Hygiene Society, 3.8, 3.74, 3.87
British Standard Specifications, and regulations, 1.49
British Standards:
  BS 1025, 4.225
  BS 1042, 4.20
  BS 2769, 4.127
  BS 4142, 3.136
  BS 4275, 3.101
  BS 4293, 4.127
  BS 5228, 3.135
  BS 5304, 4.74, 4.171
  BS 5306, 4.46, 4.51
  BS 5345, 4.124
  BS 5378, 4.58
  BS 5499, 4.58
  BS 5501, 4.124
  BS 5588, 4.56
  BS 5744, 4.101, 4.184
  BS 5750, 2.12, 2.42
  BS 5958, 4.120
  BS 5973, 4.176, 4.177, 4.179
  BS 6100, 3.165
  BS 7071, 4.127
  BS 7121, 4.184
  BS 7355, 3.98
  BS 7375, 4.123
  BS 7430, 4.116, 4.123
  BS 7671, 4.116, 4.152
  BS 7750, 2.12, 2.42
  BS 7863, 4.64
  BS CP 3010, 4.184
  BS CP 6031, 4.167, 4.168, 4.169
  BS EN 3, 4.46
  BS EN 81, 4.141
  BS EN 292, 4.77, 4.79
  BS EN 294, 3.184, 4.75, 4.90
  BS EN 349, 3.184, 4.75, 4.99
  BS EN 574, 4.89
  BS EN 614, 3.184
  BS EN 811, 4.90
  BS EN 953, 4.74
  BS EN 999, 3.184, 4.75, 4.88
  BS EN 1050, 4.78, 4.224
  BS EN 1088, 4.74, 4.80
  BS EN 1400, 2.42
  BS EN 8800, 2.42
  BS EN 9000, 2.160
  BS EN 9001, 2.12, 2.42, 2.160
  BS EN 14001, 2.12, 2.42, 2.160
  BS EN 24869, 3.99
  BS EN 45004, 4.154
  BS EN 60079, 4.124
  BS EN 60529, 4.123
  BS EN 60825, 3.120
Broadcast advertising, 1.128
Bronchial asthma, 3.40–3.41, 3.54
Building materials, and fires, 4.36–4.37
Building site safety see Construction site safety
Burden of proof, 1.21, 1.25, 1.37
Burns, electrical, 4.115
Business interruption, 2.12–2.13

Cadmium, illness and diseases from, 3.46
Cancer, 3.25–3.26
  and bladder, 3.71
  carcinogen classification, 3.56
  carcinogenicity tests, 3.30–3.31
  causes of, 3.56
  from coal tar and pitch, 3.35–3.36
  occupational, 3.55–3.57
  and pneumonia, 3.55
  and radiation, 3.108
  of scrotum, 3.34–3.35
Candela, 3.165
Capability, of employees, 1.103
Carbohydrates, 3.11
Carbon, and organic compounds, 4.5
Carbon dioxide, effects of, 3.53
Carbon disulphide, toxic effects, 3.51
Carbon monoxide, dangers of, 3.52
Carbon tetrachloride:
  harmful effects, 3.50
  and ozone depletion, 4.236
  uses and metabolism, 3.50
  ventilation for, 3.94
Carcinogens see Cancer
Cardiovascular, illnesses with causes of, 3.71
Carpal tunnel syndrome, 3.186–3.187
Carter R.L., 2.44
Case law interpretation, 1.38
Cattell's trait theory/personality factors, 2.130
Causal networks, 2.121–2.122
CE mark, 1.47
Central nervous system, 3.20
Cerebellum, 3.20
Chancery, 1.11

Change, causes and effects:
  and accident susceptibility, 2.131–2.132
  by environmental change, 2.134
  cultural factors, 2.129
  from environmental factors, 2.128–2.129
  from learned characteristics, 2.128–2.129
  from learning and training, 2.134–2.136
  genetic factors, 2.128
  goals and objectives, 2.134
  maladaptation, 2.134
  methods of, 2.133–2.134
  personality and attitudes, 2.129–2.130
  situational factors, 2.129
  stimuli and resistance to, 2.133
Chartered Institution of Building Services Engineers (CIBSE), 3.168–3.169
Chemical Industries Association (CIA), 4.231, 4.245
Chemicals:
  chemical reactions, 4.5
  Chemical Weapons Convention (CWC), 4.194
  COSHH regulations, 4.192
  emergency procedures, 4.230–4.231
  exposure monitoring, 4.201–4.202
  exposure prevention:
    enclosure, 4.199
    local exhaust ventilation (LEV), 4.199
    personal protective equipment (PPE), 4.199–4.201
    quantity minimising, 4.199
    respiratory protective equipment (RPE), 4.200
    substitution, 4.198–4.199
  hazardous substance lists, 4.192
  hazards, types of, 4.193
  immediate/long-term dangers, 2.14
  laboratories, 4.228–4.230
  legislation, 4.202–4.212
    carriage requirements, 4.204
    Coshh regulations, 4.206–4.208
    environmental protection, 4.208–4.212
    hazardous site control, 4.203–4.204
    information and packaging, 4.204–4.206
    installation notification, 4.202–4.203
    labelling, 4.204–4.206
    pollution control, 4.208–4.211
    statutory nuisances, 4.212
    supply requirements, 4.204
    waste disposal, 4.211
  new substances, 4.193–4.194
  plant/process design, 4.217–4.226
    computer control systems, 4.221–4.222
    functional safety life cycle management (FSLCM), 4.225–4.226
    Hazard and Operability Studies (HAZOP), 4.219–4.220
    modification procedures, 4.226
    plant control systems, 4.221–4.222
    risk assessment, 4.222–4.225
    safety, 4.218–4.221
  properties, 4.5–4.9
  risk, 2.26
  risk management, 4.198–4.202
  risk/hazard assessment, 4.194–4.198
    COSHH, 4.195–4.197
    definitions, 4.194
    forms for, 4.196–4.197
    manual handling, 4.197
    process of, 4.195
  safe systems of work, 4.226–4.228
    instruction documentation, 4.226–4.227
    permits to work, 4.227–4.228
    training, 4.227
  safety information, 4.193–4.194
  safety monitoring, 4.201
  safety records, 4.201
  safety training, 4.201
  storage, 4.212–4.216
    drum compounds, 4.212–4.215
    gas cylinders, 4.216
    storage tanks, 4.212–4.215
    underground tanks, 4.215
    warehousing, 4.215–4.216
  tanker off-loading, 4.215
  transport, 4.216–4.217
Chlorine, illness from, 3.52
Chlorofluorocarbons (CFCs), and ozone depletion, 4.236
Chromium, skin problems from, 3.45
Circadian rhythm, 2.110–2.111
Circuit judges, 1.23
Circulatory system, 3.16–3.18
Civil actions:
  applications, 1.5
  burden of proof for, 1.5
  defences to, 1.41
  for injury, 1.3–1.4
  time limits for, 1.41
Civil law, 1.4–1.5
Civil liability, 1.147–1.161
  claim defences, 1.156–1.158
  the common law, 1.147
  damage assessment, 1.159–1.160
  employer's liability, 1.153–1.155
  fatal accidents, 1.160–1.161
  health and safety at work, 1.156
  no fault liability, 1.161
  occupier's liability, 1.151–1.152
  supply of goods, 1.152–1.153
  time limitations, 1.158–1.159
  tort, law of, 1.148–1.151
  volenti non fit injuria, **1.158**
Clean air *see* Air pollution/sampling/control
Cleanliness of premises and equipment, 1.58
Clerk to the Justices, 1.23
Clothing, protective, 3.101–3.102

Coal tar, cancer from, 3.35–3.36
Cobalt, pneumoconiosis from, 3.37
Code for Interior Lighting (CIBSE), 3.168–3.169
Codes of practice:
  HSW definition, 1.53
  and trade descriptions, 1.119
Collective agreements, 1.82
Colour blindness, 3.164
Combustion *see* Fire(s)
Committé European de Normalisation Electrotechnique (CENELEC), 1.70, 1.71
Committee on Consumer Protection report (Cmnd 1781), 1.114
Committee on Product Safety Emergencies (EU), 1.124
Common law, the, 1.147, 2.157
Communication, face to face/written/visual, 2.136–2.138
Compensation:
  and contracts, 1.80
  and dismissal, 1.110
  for injury, 1.3
  see also Insurance cover and compensation
Competent persons, 4.153–4.154
Complete Health and Safety Evaluation (CHASE), 2.53
Complex man, 2.101–2.102
Compressed air, on construction sites, 4.176
Computers, use of:
  alphanumeric data, 2.92–2.93
  choosing software, 2.94
  free format text programs, 2.91–2.92
  hardware and software, 2.90
  numeric data, 2.94
  programs, nature of, 2.90–2.91
  question/answer programs, 2.93
Concept learning, 2.135–2.136
Conditional fee agreements, 1.22
Conditioning, and stimulus response learning, 2.135
Conjunctivitis, 3.164
Consideration, and contracts, 1.78
Constant Attendance Allowance, 1.135
Constitutional law, 1.5–1.7
Construction site safety, 4.161–4.189
  accident rates, 4.161
  asbestos legislation, 4.189
  cold and wet conditions, 4.173
  compressed air hazards, 4.176
  drinking water, 4.185
  dusty conditions, 4.174
  electrocution dangers, 4.172
  eye protection, 4.185
  fire, 4.173
  fire certificates, 4.186–4.187
  fire legislation, 4.186–4.187
  first aid boxes, 4.184–4.185
  food legislation, 4.188
  fumes, 4.174
  glare problems, 4.173
  head protection, 4.185–4.186
  hot conditions, 4.174
  industrial dermatitis, 4.174–4.175
  ionising radiations, 4.175
  lasers on, 4.176
  mines, 4.188–4.189
  non-destructive weld testing, 4.175
  notification of work, 4.164–4.165
  personal protective equipment (PPE), 4.185
  petroleum spirit legislation, 4.188
  planning supervisors, 4.162
  principal contractor responsibilities, 4.162, 4.163–4.164
  quarries, 4.188–4.189
  sewers, 4.175
  vibration white finger (VWF), 4.175
  washing facilities, 4.188
  Weil's disease, 4.175
  welfare facilities, 4.184–4.185
  see also Excavation site safety
Consumer contracts, 1.129–1.130
Consumer credit advertising, 1.120–1.121
Consumer protection:
  consumer redress, 1.130–1.131
  exclusion clauses, 1.128–1.130
  fair contract conditions, 1.114–1.121
  fair quality of goods, 1.121–1.122
  fair quality of services, 1.122
  misleading advertising, 1.127–1.128
  product liability, 1.126–1.127
  product safety, 1.122–1.124
Consumer redress, 1.130–1.131
Consumer Safety Unit (DTI), 1.126
Contaminated land, 4.237
Contracts/contract law, 1.5, 1.78–1.87
  consumer contracts, 1.129–1.130
  contract formation, 1.78–1.79
  employment, 1.39–1.40, 1.81–1.84
  fair conditions of, 1.114–1.121
  faults in, 1.79–1.80
  insurance, 1.85
  law of sale, 1.84–1.86
  and occupational safety advisers, 1.86–1.87
  remedies, 1.80–1.81
  of service and services, 1.81
  unfair contracts, 1.85
Contributory negligence, 1.146
  defence, 1.143, 1.157
Control of Substances Hazardous to Health Regulations (COSHH) *see* Appendix 6
Controlled Waste, 1.56
Cooper, M.D. *et al.*, 2.150
Corporate killing proposals, 1.22
Corporate liability, 1.37
Corrected Effective Temperature (CET) index, 3.154

Cortisone, 3.23
Cost benefit analysis, risk, 2.50
Costs, insured/uninsured, 2.18
Coughing, 3.14, 3.26
Courts:
   of Appeal, 1.9, 1.10, 1.11, 1.13, 1.14, 1.15, 1.16, 1.20, 1.23
   appellate jurisdiction, 1.7
   of Auditors (EU), 1.27, 1.69
   Civil:
      England, 1.7–1.11
      Northern Ireland, 1.14–1.16
      Scotland, 1.12–1.14
   County, 1.9
   Court of First Instance (EU), 1.27
   Court of Justice (EU), 1.26–1.27
   Criminal:
      England, 1.8, 1.9, 1.10, 1.11
      Northern Island, 1.14, 1.16
      Scotland, 1.13–1.14
   Crown, 1.10, 1.11, 1.20, 1.23
   District, 1.13
   first instance jurisdiction, 1.7
   High Court, 1.9, 1.10, 1.11, 1.14, 1.15
   High Court of Justiciary, 1.13, 1.14
   Inferior courts, 1.7–1.8
   Magistrates Courts, 1.10, 1.11, 1.14, 1.15
   personnel:
      England, 1.22–1.24
      Northern Ireland, 1.24
      Scotland, 1.24
   procedures, 1.18–1.22
   Sheriff Court, 1.13
   Superior courts, 1.7–1.8
   Supreme Court of Judicature, 1.11
   see also European courts
Cramp, from heat, 3.64
Cranes:
   access, 4.100
   bare conductors, 4.99
   checklists, 4.180–4.182
   controls, 4.99
   definition, 4.95
   emergency escape, 4.99
   hand signals, 4.100, 4.101
   inspection, 4.98
   load indicators, 4.99, 4.181
   overtravel switches, 4.99
   passengers, 4.100
   safe use, 4.99–4.100, 4.180–4.182
   safe working load, 4.100
   safety catches, 4.99
   see also Lifts/hoists/lifting equipment
Credit and hire advertising, truth in, 1.120–1.121
Crimes, definition, 1.4
Criminal cases:
   burden of proof for, 1.4
   proceedings for, 1.20
   rules of evidence for, 1.5
   where held, 1.4

Criminal law, 1.4–1.7
Criminal liability, 1.7
Criminal offences away from work, 1.106, 1.107
Criminal proceedings, 1.8
Crown immunity, 1.58–1.59
Crown Notices, 1.58
Crown premises, legislation for, 1.58–1.59
Crown Prosecution Service, 1.4
Cumulative trauma disorder (CTD), 3.186
Current Law, abbreviation list, 1.18
Customs officers, powers of, 1.51
Cyclones, 3.151

Damage:
   from negligence, 1.150–1.151
   laws of, 1.144–1.146
Damage control, 2.10
   reporting/investigating/costing, 2.46–2.47
Damages:
   assessment of, 1.159–1.160
   and contracts, 1.80
   provisional damage, 1.159
   special damage, 1.159
Danger:
   action plans, 2.127
   action plans for, 2.127
   assessment of, 2.126
   choice to expose to, 2.123–2.124
   definition, 2.5
   forseeability of, 2.125
   and personal control, 2.124
   prediction, 2.116–2.117, 2.119–2.121
   probability assessment, 2.126
   reactions to, 2.122–2.126
   reactions to perceived risk, 2.122–2.126
   responsibility for action, 2.126–2.127
   see also Hazards
Dangerous goods, carriage of, 1.57
Dangerous occurrences, UK notification requirements, 2.65–2.66
de Quervain's disease, 3.187
Deafness, occupational, 3.64, 3.190
Debris, fire danger from, 4.33
Decibel, 3.125–3.126
Decision making and intelligence, 2.107–2.108
Defences, to civil action, 1.41, 1.156–1.158
Dehydration, 3.64
Delegated legislation, 1.29–1.30
Deoxyribonucleic acid (DNA), 3.107–3.108
Dermatitis, non-infective, 3.33–3.34
Design:
   health and safety in, 2.40–2.42
   influence on behaviour, 2.138–2.139
   for safety, 2.116
Development risks defence, 1.127
Digestive system, 3.12–3.13
Directives, European Union (EU), 1.32–1.34
Director disqualification, 1.36

Appendices

Director of Public Prosecutions (DPP), 1.4, 1.24
Disabilities:
  disability discrimination laws, 1.95–1.96, 3.190, 3.193
  and ergonomics, 3.190–3.191
Disabled people, fire escape, 4.56
Disciplinary procedures, 1.99–1.101
  ACAS Code of Practice, 1.99
Disciplinary rules, 1.82, 1.100
Disclaimers, and false descriptions, 1.118
Discrimination *see* Disabilities; Race discrimination; Sex discrimination; Young persons
Diseases:
  de Quervain's disease, 3.187
  Legionnaire's disease, 3.67
  from metals, 3.42–3.46
  from micro-organisms, 3.65–3.69
  from pesticides, 3.46–3.47
  respiratory, 3.36–3.42
  skin, 3.33–3.36, 3.70
  from solvents, 3.47–3.51
  UK notification requirements, 2.68
  Weil's disease, 4.175
  see also occupational diseases
Dismissal, 1.102–1.105
  and capability of employees, 1.103
  and compensation, 1.110
  and continuing absences, 1.104
  for contravention of an enactment, 1.109
  effective date of termination, 1.102
  fair/unfair reasons, 1.83, 1.102–1.103, 1.109–1.110
  and ill-health, 1.103–1.104
  for lack of skill, 1.104–1.105
Display screens, 3.180–3.181
  legal requirements, 3.192
Diurnal rhythm, 2.110–2.111
Domestic premises, HSW definition, 1.53
Domino theory, 2.85–2.88
Domino theory of accidents, 2.14–2.15
Doors:
  for fire escape, 4.57
  and fire prevention, 4.39
Double vision, 3.164
Douglas H.M., 2.55
Draeger tube, 3.153
Due diligence defence, false trade descriptions, 1.116
DuPont Safety Training Observation Programme (STOP), 2.152
Dust:
  air pollution/sampling/control, 3.74, 3.81, 3.82, 3.83–3.84, 3.96, 3.150–3.152
  airborne dust lamps, 3.83–3.84
  asbestos dust control, 4.150–4.151
  fire danger from, 4.35
  flammable, 4.28, 4.29

Duty:
  practicable, 1.37
  standards of, 1.37–1.38
Duty of care, 1.38–1.39
  by employers, 1.39–1.40
  environmental, 4.241
  and negligence, 1.149–1.150

Ear-muffs, 3.98–3.99, 3.144
Ear-plugs, 3.98–3.99, 3.144
Ears, 3.21–3.22
  inner ear, 3.21–3.22
  working of, 3.133
  see also Hearing
Eczema, 3.34
Electricity:
  accidents, 4.111–4.112
  in adverse/hazardous environments, 4.123
  alternating current, 4.107
  burns from, 4.115
  cable fires, 4.39
  circuit breakers, 4.118
  circuits, 4.112–4.113
  competency for working, 4.118–4.119
  construction site fires, 4.173
  construction site safety, 4.172
  direct current, 4.107–4.108
  earth leakage circuit breaker, 4.118, 4.119
  earthing, 4.116
  electric discharge lamps, 3.166
  electric fields, 3.121–3.122
  electric shock, 3.18, 4.113–4.115
  in explosive atmospheres, 4.123–4.124
  as a fire danger, 4.31
  fires from, 4.30, 4.115–4.116
  fittings, 4.125
  flameproof equipment, 4.126
  in flammable atmospheres, 4.123–4.124
  fuses, 4.117–4.118
  hazardous areas classification, 4.124–4.125
  impedance, 4.113
  inspection and test of equipment, 4.152–4.153
  insulation, 4.117
  intrinsically safe systems, 4.125
  maintenance of equipment, 4.127–4.128
  Ohm's law, 4.112
  overhead lines, HSE notes, 4.118
  permits to work (PTW), 4.116, 4.120, 4.121–4.122
  personal protective equipment (PPE), 4.117
  portable tools, 4.126–4.127
  pressurising, 4.125
  purging, 4.125
  residual current devices (RCDs), 4.127
  static, 4.120, 4.123
  statutory requirements, 4.110–4.111

supply, 4.108–4.110
type 'n' and 'e' equipment, 4.125
underground cables, HSE notes, 4.118
voltage, 4.108, 4.111
work precautions, 4.116–4.117
Electrons, 3.105–3.106, 4.3–4.4
Emissions, control of, 1.46
Employee, HSW definition, 1.53
Employee dismissal *see* Dismissal
Employee responsibility for H&S, 74 and legislation, 1.82
Employee rights, 1.91–1.92
Employee suspension, 1.82–1.83
Employers:
  and blatant disregard of H&S proposals, 1.22
  civil liability of, 1.153–1.156
    for fellow employee selection, 1.155
    for place of work, 1.154
    for plant and equipment, 1.154–1.155, 1.156
    for supervision/instruction, 1.155
    for system of work, 1.153–1.154
    and vicarious liability, 1.153
  duty of care, 1.39–1.40, 1.46
  duty legislation, 1.82, 1.138
  general duties on, 1.46–1.48
  HSW requirements, 1.49–1.50
  liability insurance, 1.136–1.141
  responsibility to observe legislation, 1.45
Employment:
  contracts for, 1.81
  documents and agreements, 1.92
  Employment Appeal Tribunal, 1.25
  Employment Medical Advisory Service, 1.53
  law, 1.90–1.92
  legislation, 1.81–1.84
  see also Dismissal
Encephalopathy, 3.43
Endothermic reactions, 4.5
Energy, work and power, 4.15–4.16
Enforcement notices, environmental, 4.240–4.241
Engines, as a fire danger, 4.31–4.32
Environment, change to, 2.134
Environmental Agency (EA), 4.244
Environmental issues/legislation, 4.234–4.238, 4.238–4.245
  acid rain, 4.236
  air pollution control (APC), 4.240
  Basle Convention, 4.238–4.239
  chemicals, protection from, 4.208–4.212
  clean air legislation, 4.242–4.243
  contaminated land, 4.237
  controlled waste legislation, 1.56, 4.242
  duty of care legislation, 4.241
  Environmental Agencies, 1.56
  global warming, 4.234–4.235
  Inspectorate of Pollution, 1.56
  integrated pollution control (IPC), 4.240

Montreal Protocol, 4.240, 4.238
noise legislation, 4.243
noise and nuisance, 4.238
ozone depletion, 4.235–4.236
photochemical smog, 4.236
protection legislation, 4.239–4.241
resource depletion, 4.238
special waste legislation, 4.243–4.244
waste classification, 4.239
waste disposal, 4.237, 4.240
waste management legislation, 4.243–4.244
water pollution, 4.237
water resources legislation, 4.241–4.242
see also Pollution
Environmental management systems, 4.245–4.247
  BS EN ISO 1400 series of standards, 4.245–4.247
  objectives, 4.247
Epicondylitis, 3.186
Epidemiological analysis:
  with limited data, 2.79
  purpose of, 2.78
  single/multi-dimensional, 2.78–2.79
  techniques for, 2.78–2.79
Epidemiology, 3.32–3.33
Equal opportunities *see* Disabilities; Race discrimination; Sex discrimination; Young persons
Equal pay legislation, 1.96
Equipment:
  food, cleanliness, 1.58
  see also Plant
Ergonomics, 3.176–3.194
  allocation of function, 3.183
  anthropometry, 3.183–3.184
  and back problems, 3.188–3.190
  by design, 3.179
  controls, 3.182–3.183
  definition, 3.176–3.177
  and the disabled, 3.190–3.191
  displays, 3.180–3.181
  and error, 3.184–3.185
  history of, 3.177–3.179
  human-machine interface, 3.180
  legal requirements, 3.191–3.193
  text clarity, 3.181–3.182
  unstability, 3.179–3.180
  and work related upper limb disorders (WRULD), 3.185–3.188
  in the workplace, 3.185
Errors:
  causes of, 2.112–2.116, 2.121
  and ergonomics, 3.184–3.185
  quantification of, 2.121
Erythema, 3.119
European Agency for Health and Safety at Work, 1.33, 1.69
European Committee for Electrotechnical Standardisation (CENELEC), 4.124

European courts, 1.9, 1.11, 1.12, 1.15
  European community courts (ECJ),
    1.25–1.28
  European Court of Human Rights,
    1.27–1.28
  European Court of Justice, and judicial
    precedent, 1.18
  European Union (EU):
    Court of Auditors, 1.27, 1.69
    Court of Justice, 1.69
European Parliament, 1.33, 1.68, 1.69
European standards, 1.70–1.71
European Union (EU):
  Commission, 1.68
  Committee of Permanent
    Representatives, 1.68
  Council of, 1.67
  Decisions and Recommendations of,
    1.69
  Directives, 1.32–1.34, 1.69
  Economic and Social Committee
    (EcoSoc), 1.68
  and employer employee relationships,
    1.71–1.72
  influence on HSW legislation, 1.36–1.37
  legislation, 1.31–1.33, 1.69–1.70
    applied to individuals, 1.33
  machinery directives, 4.69–4.73
  and product information exchange,
    1.123–1.124
  Regulations of, 1.69
Examination of plant and equipment:
  boilers, 4.134–4.137
  competent persons, 4.153–4.154
  cranes and lifting machines, 4.141–4.144
  defects, latent and patent, 4.154–4.155
  electrical equipment and installations,
    4.152–4.153
  gas containers, 4.131–4.134
  history of, 4.130–4.131
  legislation, 4.131
  lifting and handling equipment,
    4.139–4.148
    operations regulations, 4.147–4.148
    records of tests regulations,
      4.146–4.147
  lifting tackle, 4.145–4.146
  lifts and hoists, 4.139–4.141
  power presses and press brakes,
    4.148–4.149
  pressure systems, 4.131–4.139
  process of, 4.155
  reporting, statutory requirements,
    4.155–4.156
  steam and air receivers, 4.137–4.139
  ventilation, 4.149–4.151
Excavation site safety, 4.165–4.171
  battering the sides, 4.165, 4.166
  benching the sides, 4.165, 4.166
  flooding risks, 4.166
  guardrails and barriers, 4.166

mobile machinery on, 4.167–4.168, 4.170,
  4.171
overhead cables, 4.168
portable tools, 4.171
pre-planning, 4.170–4.171
public access, 4.167
steel trench sheet runners, 4.169
underground cables, 4.168–4.170
walling poling frames, 4.168
see also Construction site safety
Exchange and barter, 1.84
Exclusion clauses, 1.128–1.130
Excretion, 3.13
Expert witnesses/evidence, 1.22
Explosive material, 1.55
Exposure limits, air contamination *see* Air
  pollution/sampling/control
Extinction, management technique, 2.145
Extinguishers *see* Firefighting
Extremely low frequency fields (ELF),
  3.121–3.122
Extrinsic allergic alveolitis, 3.41–3.42
Eyes, 3.20–3.21
  burns of, 3.21
  cataracts, 3.21
  conditions/disorders, 3.164–3.165
  eye strain, 3.164
  illnesses with causes of, 3.70
  protection, 3.101
  protection on construction sites, 4.185

Facilities, and false descriptions,
  1.116–1.117
Factories, early safety legislation,
  1.34–1.35
Facts, and the law, 1.7
Failure modes, accidents, 2.16
Fairbairn, Sir W., 4.130
False trade descriptions:
  code of practice for, 1.119
  disclaimers, 1.118
  due diligence defence, 1.116, 1.119
  list of descriptions, 1.115–1.116
  penalties, 1.118, 1.120
  price indications, 1.120
  pricing offences, 1.118–1.120
  reckless statements, 1.117–1.118
  services, facilities and accommodation,
    1.116–1.118
  strict liability offence, 1.116
  truth in lending, 1.120–1.121
Farmer's lung, 3.41–3.42
Fatal accidents, 1.160–1.161
  fatal accident frequency rates (FAFRs),
    2.39–2.40
Fatalities, UK notification requirements,
  2.67
Fatigue, muscular and mental, 2.109–2.110
Fault tree analysis:
  accidents, 2.16
  risk, 2.37–2.38

Feedback, of discussions, 2.22
Fibres, sampling for, 3.81, 3.83–3.84
Financial accountability, for risk management, 2.51
Fire(s):
  alarms, 4.40–4.41
  and building materials, 4.36–4.37
  chemistry of, 4.23–4.27
  class A, solid organic materials, 4.29
  class B, liquids and liquifiable solids, 4.29
  class C, gases, 4.30
  class D, flammable metals, 4.30
  combustion process, 4.27–4.29
  combustion products, 4.35
  on construction sites, 4.173
  construction sites, 4.186–4.187
  detectors, 4.41–4.42
  dusts, flammable, 4.28, 4.29
  electrical, 4.30, 4.115–4.116
  extinction methods, 4.42–4.46
    cooling, 4.45
    fuel starvation, 4.43–4.44
    with halons, 4.45–4.46
    smothering, 4.44–4.45
  extinguishing agents, 4.30
  Fire point temperature, 4.26
  Fire triangle, 4.24–4.25
  flameproof electrical equipment, 4.126
  flammable substances, 4.27
  flash point temperature, 4.26
  fuel for, 4.25
  ignition/ignition temperature, 4.25, 4.26
  liquids, flammable, 4.28
  lower flammable (explosive) limit, 4.26
  and oxygen, 4.25
  risk levels, 4.55
  smoke, 4.28
  solids, flammable, 4.28
  spontaneous combustion, 4.28–4.29
  spread of, 4.35–4.36
  structural precautions, 4.38–4.40
  upper flammable (explosive) limit, 4.26
  see also Fire escape; Fire legal requirements; Firefighting; Flammable substances, control and storage; Ignition sources
Fire Authority, 1.55
Fire certificates, 1.54–1.56
  construction sites, 4.186–4.187
  improvement and prohibition notices, 1.55
Fire escape:
  disabled people, 4.56
  doors, 4.57–4.58
  emergency plans, 4.58–4.59
  exits, 4.56
  gangways, 4.57
  instruction and training, 4.59–4.60
  lighting, 4.58
  maintenance and record keeping, 4.60
  notices, 4.58
  panic bolts and latches, 4.57–4.58
  passages, 4.57
  principals of, 4.55–4.56
  signs, 4.58
  stairways, 4.57
  travel distances, 4.56, 4.57
Fire legal requirements, 1.54–1.56, 4.60–4.64
  appeals and offences, 4.62
  designated premises, 4.61
  fire authority duties, 4.60
  fire authority powers, 4.62–4.63
  fire certificates, 4.61–4.63
  regulations, 1.50
  risk assessments, 4.63
Fire precautions/prevention, 4.23–4.67
  loss control, 2.10–2.11
Firefighting, 4.46–4.55
  fixed equipment:
    automatic sprinklers, 4.52–4.55
    carbon dioxide systems, 4.54
    drencher systems, 4.54
    dry pipe sprinklers, 4.53
    fixed foam systems, 4.54
    halon systems, 4.54
    hose reels, 4.51
    pre-action sprinklers, 4.54
    rising mains, 4.51–4.52
    wet pipe sprinklers, 4.53
  liaison with fire brigade, 4.64
  portable extinguishers, 4.46–4.51
    carbon dioxide type, 4.48
    dry powder type, 4.49
    foam type, 4.49
    operation, 4.46–4.50
    provision, location and maintenance, 4.50–4.51
    water type, 4.47, 4.48
First aid boxes, 3.10
  construction sites, 4.184–4.185
First aid/first aiders at work, 3.9–3.10
First instance jurisdiction, 1.7
Fitness for purpose, 1.121–1.122
Fitts, P.M., 3.183
Flame detectors, 4.42
Flameproof electrical equipment, 4.126
Flammable material, 1.55
Flammable substances, control and storage
  dusts, 4.34–4.35
  gases, 4.33–4.35
  liquids, 4.34–4.35
  waste, debris and spillage, 4.33
Fletcher, J.A., 2.55
Floors, and fire prevention, 4.39
Fluids:
  compressible *see* Gases
  incompressible *see* Liquids
Fluorescent lights, 3.167
Fog Index, Gunning, 3.181–3.182

Food:
  building site legislation, 4.188
  safety regulations, 1.126
Foodstuffs, 3.11–3.12
Force, work and power, 4.15–4.16
Forfeiture Orders, goods, 1.125
Forseeability of danger, 2.125
Frequency rates, 2.75
Freudian psychology, 2.130
Friction:
  as a fire danger, 4.31
  static and sliding, 4.16–4.17
Fuel, 4.25
Fumes, noxious or offensive, 1.47
Functional safety life cycle management (FSLCM), chemical plants, 4.225–4.226

Gamma-rays, 3.107
Ganglions, 3.187
Gangways, for fire escape, 4.57
Gas chromatography, 3.80
Gas containers:
  regulations for, 4.100–4.102
  transportable
    definition, 4.102
    safe usage, 4.103–4.104
Gas cylinders, storage, 4.216
Gases:
  compression of, 4.20
  fire danger from, 4.33–4.35
  flow measurement, 4.20–4.21
  General Gas Law, 4.11
  physical properties, 4.11–4.12
  sampling for *see* Air pollution/sampling/control
  specific gravity, 4.12
Gassing:
  accidents from, 3.51–3.53
  see also named gases
Germain, G.L., 2.10
Gilbreth, F.B, 3.178
Glare, 3.170–3.171
  glare index, 3.171
  protection from, 3.171
Glendon, A.I., 2.98, 2.116
Global warming, 4.234–4.235
Gloves:
  neoprene, 3.101
  polyvinyl alcohol, 3.102
Goal setting, 1.66–1.67
Goods:
  fair quality, 1.121–1.122
  forfeiture orders, 1.125
  notices to warn, 1.125
  and product liability, 1.126–1.127
  Prohibition Notices, 1.125
  Suspension Notices, 1.125
Gordon, J.E., 2.16
Grab sampling, 3.75–3.78

Green Papers, 1.31
Greenhouse effect, 4.235
Greenhow, Dr E.H., 3.4
Grievance procedure, 1.82
Gross misconduct, 1.101
Guards and interlocks, 3.191
  adjustable guards, 4.80
  distance guards, 4.79
  fixed enclosing guards, 4.80
  fixed guards, 4.79
  hazard identification, 4.76–4.77
  interlocked guards:
    automatic, 4.88
    cam-operated, 4.81–4.85
    captive key, 4.85
    control guards, 4.89
    direct manual, 4.81
    features and choice, 4.80–4.81
    limit switches, 4.83–4.85
    magnetic switch, 4.86
    mechanical, 4.81, 4.82
    mechanical scotches, 4.86–4.87
    time delayed, 4.86–4.87
    trapped key, 4.83–4.85
    trip devices, 4.88
    two-hand control, 4.88
  materials for, 4.89–4.90
  openings, 4.90
  reaching over, 4.90
  risk assessment, 4.77–4.78
  selection and monitoring, 4.77–4.78
  tunnel guards, 4.79
Guilty plea, 1.21
Gunning Fog Index, 3.181–3.182

Hale, A.R., 2.84, 2.98, 2.116
Hale, M., 2.84
Halons, and ozone depletion, 4.236
Hand-arm vibration syndrome (HAVS), 3.57–3.58
Hands/arms, illnesses with causes of, 3.70
Hazards:
  definition, 2.4
  degrees of, 2.13–2.14
  hazard analysis (HAZAN), 2.39
  hazard and operability (HAZOP) studies/techniques, 2.23, 2.38, 2.119–2.120
  chemical plant, 4.219–4.220
  Hazard and Risk Explained HSE leaflet, 2.4–2.5
  inspections for, 2.119
  prediction techniques, 2.119–2.121
  and risk, 2.4–2.5
  routine detection, 2.118–2.119
  warnings of, 2.118–2.119
  see also Danger; Health hazards; Risk
Head protection, construction sites, 4.185–4.186

Health, dismissal for ill-health,
  1.103–1.104
Health assessments, 3.6
Health hazards, classes of, 3.75
Health promotion, 2.124–2.125
Health and safety, in design and planning,
  2.40–2.42
Health and safety at work legislation,
  1.44–1.59
  administration of 1974 act, 1.48–1.49
  definitions of 1974 act, 1.53–1.54
  Employment Medical Advisory Service,
    3.194
  enforcement, 1.50–1.51
  extensions to act, 1.52
  formal risk assessment, 2.161
  framework for, 1.64–1.65
  general duties on employers and
    others, 1.46–1.48
  goal setting for, 1.66–1.67
  offences, 1.51–1.52
  post 1974, 1.45–1.46
  pre 1974, 1.44–1.45
  regulations and codes of practice,
    1.49–1.50
  see also Construction site safety;
    Occupational health, hygiene and
    safety
Health and Safety Commission (HSC),
  1.46, 1.63–1.64
  duties of, 1.48
Health and Safety Executive (HSE)
  duties of, 1.48–1.49
  Occupational Exposure Limits (OEL),
    3.155
  operation of, 1.63–1.64
Hearing:
  audiogram, 3.63, 3.133
  deafness and noise, 3.70
  ear sensitivity, 3.61
  hearing protectors, 3.99
  mechanism of, 3.61
  noise exposure limits, 3.62–3.63
  noise induced loss claims, 1.140–1.141
  occupational deafness, 3.64, 3.134
  protection for, 3.63, 3.98–3.99, 3.137
  titnnitus, 3.134
  see also Ear
Heart, 3.16–3.18
Heat, as a form of energy, 4.15
  Heat detectors, 4.41–4.42
Heat Stress Index (HSI), 3.154
Heat stroke, 3.64
Heated environment see Thermal
  environment
Heating systems, as a fire danger, 4.31
Heinrich, H.W., 2.14, 2.15, 2.22
Hepatitis A/B/C, 3.65–3.66
Herbicides, dangers of, 3.47
Hire and credit advertising, truth in,
  1.120–1.121

Hire-purchase agreements, 1.87
HIV +ve, 3.66
Hoists see Lifts/hoists/lifting equipment
Hooke's Laws, 4.17
Hormones, 3.23–3.24
Houldsworth, H., 4.130
House of Lords, 1.8, 1.9, 1.10, 1.11, 1.12,
  1.14
Human body, structure and functions,
  3.3–3.28
  anatomy and physiology, 3.10–3.25
  cancer, 3.25–3.26
  central nervous system, 3.20
  circulatory system, 3.16–3.18
  defence mechanisms, 3.26–3.27
  digestion, 3.12–3.13
  ear, 3.21–3.22
  excretion, 3.13
  eye, 3.20–3.21
  heart, 3.18
  history, 3.3–3.5
  hormones, 3.23–3.24
  muscles, 3.19
  poisons, effects of, 3.27–3.28
  respiratory system, 3.13–3.16
  skin, 3.24–3.25
  smell and taste, 3.22–3.23
  special senses, 3.20
Human factor and health and safety,
  2.96–2.142
  behaviour modification, 2.144–2.152
  human being as a system, 2.98–2.99
  and performance management,
    2.145–2.150
Human function and behaviour
  aptitude tests, 2.107
  attitude, 2.132–2.133, 2.143
  behaviour intervention process,
    2.149–2.150
  behaviour observation and counselling,
    2.151–2.152
  behavioural analysis, 2.147
  behavioural process, future for,
    2.152–2.155
  and communication, 2.136–2.138
  complex man, 2.101–2.102
  decision making and intelligence,
    2.107–2.108
  design, influence of, 2.138–2.139
  diurnal and circadian rhythm,
    2.110–2.111
  environmental effects, 2.108–2.109
  error quantification, 2.121
  errors, causes of, 2.112–2.116, 2.121
  expectancy, 2.104–2.105
  fatigue, muscular and mental,
    2.109–2.110
  goals, objectives and motivation,
    2.100–2.102
  human actions, 2.108
  intelligence quotient (IQ), 2.107

Human function and behaviour – *continued*
  Maslow A.H. and Self Actualising Man, 2.101
  Mayo E. and Social man, 2.100–2.101
  memory, short/long-term, 2.105–2.106
  performance degradation causes, 2.109–2.110
  routines and skills, 2.106–2.107
  sense organs, 2.102–2.103
  stress, 2.111–2.112, 3.23, 3.69–3.70, 3.187
  switching/attention mechanisms, 2.103–2.104
  Taylor F.W. and Economic Man, 2.100
  see also Behavioural science; Change; Danger
Humidifier fever, 3.68
Hydraulics, 4.20–4.21
  hydraulic power transmission, 4.13–4.14
Hydrochloric acid, illness from, 3.52
Hydrogen sulphide, effects of, 3.53
Hygiene standards, 3.86–3.87
  see also Air pollution/sampling/control
Hygrometry/hygrometers, 3.85–3.86, 3.154
Hypermetropia, 3.164
Hyperplasia, 3.25

Ignition sources:
  electricity, 4.31
  engines, 4.31
  friction, 4.31
  heating systems, 4.31
  hot surfaces, 4.31
  smoking, 4.31
  tools, 4.31
Ignition/ignition temperature, 4.25, 4.26
Ill health, and dismissal, 1.103–1.104
Illuminance, 3.165
  see also Lighting
Immune system, body, 3.26–3.27
Impairment assessment, 1.134
Implied terms/conditions
  in contracts, 1.83–1.84
  in sales, 1.84–1.86
Improvement Notices, 1.51
  fire, 1.55
Incandescent lamps, 3.165, 3.167
Incapacity benefit, 1.135
Incident, at Hazards Ltd, 1.3–1.4, 1.5, 1.20
Incident recall technique, and loss control, 2.10
Incidents, 2.3–2.4
  and accidents, 2.3–2.4
  definition, 2.57
  incidence rates, 2.74
  possible legal proceedings from, 1.18–1.21
  UK notification requirements, 2.65–2.69
  see also Accidents
Independent audits, 2.22–2.23
Induction lamp, 3.166, 3.167

Industrial dermatitis, on construction sites, 4.174–4.175
Industrial Injuries Disablement Benefit, 1.134
Industrial relations law, 1.7, 1.90–1.112
  disciplinary procedures, 1.99–1.101
  discrimination, 1.92–1.99
  dismissal, 1.102–1.105
  and employment law, 1.90–1.92
  and enactment contravention, 1.109
  misconduct, 1.105–1.107
  redundancy, 1.107–1.109
  unfair dismissal, 1.109–1.112
Industrial Tribunals, 1.24–1.25
  and accidents, 1.4
  and Improvement/Prohibition notices, 1.51
Inevitable accidents, 1.139
Information exchange on products, EU requirements, 1.123–1.124
Infrared radiation, 3.120
Injuries, UK notification requirements, 2.65–2.69
Injury prevention, 2.10
Inorganic compounds, 4.6–4.7
Insecticides, illnesses from, 3.46–3.47
Inspection, for hazards, 2.119
Inspections techniques, for reports, 2.63
Inspectorate of Pollution, 1.56
Inspectors, 1.18
  of accidents, 1.3
  powers of, 1.50–1.51
Instruction, employer's liability for, 1.154–1.155
Insulin, 3.23
Insurance contracts, 1.85, 1.85–1.86
Insurance cover and compensation, 1.133–1.146
  Employers liability insurance, 1.136–1.141
  investigation, negotiation and the quantum of damage, 1.143–1.146
  Public Liability insurance, 1.142–1.143
  State insurance scheme, 1.133–1.136
  Workmen's Compensation scheme, 1.133–1.136
Integrated Pollution Control (IPC), 1.56, 4.240
Intelligence and decision making, 2.107–2.108
Intelligence quotient (IQ), 2.107
Intention, and contracts, 1.78–1.79
International Atomic Energy Agency, 3.118
International Commission on Radiological Protection (ICRP), 3.109, 3.110, 3.118
International Safety Rating System, 2.169
International Safety Rating System (ISRS), 2.53
Interviews, for reports, 2.63, 2.81–2.84
Investigation, for insurance, 1.143–1.146
Ionising radiation *see* Radiation

Jastrzebowski, W., 3.179
Job safety (hazard) analysis, 2.23,
    2.148–2.150, 2.174–2.175
  analysis chart, 2.27–2.28
  job breakdown, 2.29–2.30
  procedure, 2.26–2.30
  safe systems of work, 2.31
  safety instructions, 2.30
Joint consultation, 1.98–1.99
Joint tortfeasors, 1.157–1.158
Judges, 1.22–1.23
Judicial Committee of the Privy Council, 1.17
Judicial precedent, 1.17–1.18, 1.147
  and European Union, 1.26
Justices of the Peace, 1.23

Kidneys, toxicity and infection, 3.71
Kirwan, B., 2.121
Knowledge-based work, errors during, 2.115–2.116
Komaki, J. *et al.*, 2.145
Krause, T.R. *et al.*, 2.145

Lamps *see* Lighting
Land-fill:
  legislation for, 4.244
  waste disposal, 4.211
Landlords, duties of, 1.47
Laser nephelometer, 3.84
Laser radiation, 3.121
Lasers, on construction sites, 4.176
Law, sources of, 1.28
Law making, power delegation for, 1.63–1.64
Lead, diseases from:
  biological monitoring, 3.43
  fume or dust hazard, 3.42, 4.151
  inorganic lead, 3.42–3.43
  organic lead, 3.43
Learning, psychological principles, 2.134–2.136
Legal aid, targeting proposals, 1.22
Legal reform proposals, 1.22
Legge, T., 3.4–3.5
Legionnaire's disease, 3.67
Legislation:
  Acts of Parliament, 1.29–1.30
  adopted, 1.31–1.33
  asbestos, 4.189, 4.202–4.212, 4.245
  Crown premises, 1.58–1.59
  enforcement, 1.50–1.51
  equal pay, 1.96
  European Union (EU), 1.31–1.34
  mines, 1.35, 1.56, 4.188–4.189
  noise, 4.243
  petroleum spirit, 1.58
  pollution, 1.56
  quarries, 1.35, 1.56
  statutory interpretation, 1.30–1.31
  water, 4.241–4.242

White/Green Papers, 1.31
  see also Chemicals, legislation; Environmental issues/legislation; Health and safety at work legislation; Safety legislation; Waste legislation
Lending, truth when, 1.120–1.121
Leptospirosis, 3.67
Liability, 1.126–1.127
  absolute and strict, 1.37
  corporate, 1.37
  see also Product liability; Public liability; Third party liability
Lifts/hoists/lifting equipment, 4.95–4.100, 4.139–4.148
  adequate strength, 4.155
  approved codes of practice, 4.98
  checklists, 4.182–4.184
  definitions, 4.95, 4.140, 4.141
  dock regulations, 4.141
  examinations/inspections/reports, 4.98, 4.140, 4.142–4.144, 4.180
  fishing vessel regulations, 4.141
  in-service requirements, 4.140
  load radius indicators, 4.143
  miscellaneous equipment, 4.145–4.146
    acceptable damage, 4.146
  passenger, safety, 4.141
  patent and latent defects, 4.154–4.155
  record requirements, 4.146–4.147
  regulations, 4.96–4.97, 4.139, 4.147–4.148, 4.178–4.184
  reports, 4.155–4.156
  safe use, 4.97–4.99
  safe working loads (SWL), 4.143
  sound material, 4.155
  types of, 4.139
  working load limit (SWL), 4.143
  see also Cranes
Liftshafts, and fire prevention, 4.39
Lighting, 3.163–3.174
  Code for Interior Lighting, by CIBSE, 3.168–3.169
  colour, 3.172–3.173
  for fire escape, 4.58
  glare, 3.170–3.171
  light meters, 3.173–3.174
  maintenance of, 3.169
  quality, 3.170
  shadow, 3.171
  stroboscopic effect, 3.171–3.172
  types of, 3.166, 3.167–3.168
Liquids:
  flammable, 4.28, 4.34–4.35
  heats of vaporisation and fusion, 4.13
  in hydraulic power transmission, 4.13–4.14, 4.21
  physical properties, 4.12–4.14
Liver:
  hepatitis and cancer, 3.71
  poison damage, 3.13

Local authorities, powers of for public health, **1.57–1.58**
Local exhaust ventilation (LEV), 3.95, 3.149, 3.156, 4.149–4.151
  asbestos dust control, 4.150–4.151
  chemicals, 4.199
  components of, 4.149–4.150
  examination for need of, 4.150
  purpose, 4.149
  Tyndall (dust) lamp, 4.151
Local Procurator-Fiscal, 1.20
Loftus, R.G., 2.9, 2.10, 2.15, 2.32, 2.38
Lord Advocate, 1.4, 1.24
Lord Chancellor, 1.23, 1.24
Lord Justices of Appeal, 1.23
Los Alfaques Camping Site disaster, 4.13
Loss control:
  and business interruption, 2.12–2.13
  damage control, 2.10
  fire prevention, 2.10–2.11
  incident recall, 2.10
  injury prevention, 2.10
  management in practice, 2.13
  and occupational health and hygiene, 2.11–2.12
  and pollution control/environmental protection, 2.12
  principle of, 2.9–2.10
  and product liability, 2.12
  profiling, 2.53–2.54
  security, 2.11
Losses, pecuniary/non-pecuniary, 1.145–1.146
Lower flammable (explosive) limit, 4.26
Lumen, 3.165
Luminaire, 3.166–3.167
Luminance, 3.165
Lungs, 3.13–3.16
  illnesses of, 3.70

Maastricht Treaty, 1.26, 1.33
McClelland, D., 2.101
Machinery, safe use of
  existing machinery, regulations, 4.73–4.76
  hazard identification, 4.76–4.77
  hazard reduction, 4.77
  legislative arrangements, 4.68–4.69
  Machinery Directives, 4.68–4.69, 4.69
  new machinery, regulations, 4.69–4.73
  operator training, 4.78
  residual risk assessment, 4.77–4.78
  see also Guards and interlocks; Lifts/hoists/lifting equipment; Powered trucks; Pressure systems
Magnetic fields, 3.121–3.122
Maintenance systems, and risk management, 2.45–2.46
Maladaptation, 2.134
Malt worker's lung, 3.42

Man-made mineral fibres, pneumoconiosis from, 3.37
Management control, lack of, 2.87–2.88
Management/worker discussions, 2.22
Manchester Steam Users' Association, 4.130
Manganese, diseases from, 3.46
Manometers, 3.161, 4.10
Manufacturers, civil liability of, 1.152
Maslow, A.H., 2.51
  and Self Actualising Man, 2.101
Mason hygrometer, 3.86
Materials:
  failures in, 4.19
  strength of, 4.17–4.19
Matter:
  changes of state, 4.11
  structure, 4.3–4.5
Maximum Exposure Limits (MELs), 3.155, 4.150
  air sampling, 3.87–3.88
  Trichloroethylene, 3.50
Maximum potential loss (MPL), 2.24–2.25
Mayo, E. and Social man, 2.100–2.101
Mechanics, 4.16–4.17
Medical grounds suspension, 1.82
Medicines, safety regulations, 1.126
Memory, short/long-term, 2.105–2.106
Mens rea, 1.37
Merchantable quality, 1.121
Merchantable/unmerchantable goods, 1.85
Mercury lamps, 3.166
Mercury, poisoning and diseases from, 3.44–3.45
  health surveillance for, 3.44–3.45
Mesothelioma, 3.40
Metal halide lamp, 3.166, 3.167
Metals:
  diseases from, 3.42–3.46
  metal fume fever, 3.45
  properties, 4.5–4.6
  see also named metals
Metaplasia, 3.25
Micro-organisms:
  diseases from, 3.65–3.69
  see also named diseases from
Mineral fibres, and pneumoconiosis, 3.37
Mines:
  early safety legislation, 1.35
  legislation, 1.56, 4.188–4.189
Misconduct, 1.105–1.107
Misrepresentation, and contracts, 1.79
Mistake, and contracts, 1.79
Mobile towers, 4.178, 4.179
Mobility Allowance, 1.135
Molecular formulae, 4.4–4.5
Monday morning fever, 3.68
Montreal Protocol 1987, 4.238
Motivational theory, 2.51–2.52
Multiple accident causation theory, 2.15–2.16

Munitions factories, 3.5
Murrell, K.F.H., 3.179
Muscle cramp, from heat, 3.64
Muscles, 3.19
Mutagenesis, 3.26
Myopia, 3.164

National Disability Council, 1.95
National Insurance schemes *see* State insurance schemes
National Radiological Protection Board (NRPB), 3.118, 3.119
Negative reinforcement, management technique, 2.145
Negligence, 1.39, 1.40
  and breach of duty, 1.150
  contributory, 1.146, 1.157
  and duty of care, 1.149–1.150
  and resultant damage, 1.150–1.151
Negotiation, for insurance, 1.143–1.146
Neoplasia, 3.26
Nerves, illnesses with causes of, 3.71
Neutrons, 3.105, 3.107, 4.3–4.4
Nickel and nickel carbonyl, illness and diseases from, 3.46
Night working, 2.111
Nitrous fumes, effects of, 3.53
No fault liability, 1.76, 1.161
Noise, 3.124–3.145
  absorbent materials, 3.140
  absorption coefficients, 3.140
  absorption treatment, 3.142–3.143
  acoustic enclosures, 3.140
  community noise levels, 3.135–3.136
  control techniques, 3.137–3.145
  damping, 3.142
  dosimeter, 3.135, 3.136
  enclosure protection, 3.138–3.140
  environmental legislation, 4.243
  equivalent noise level, 3.134–3.135
  lagging, 3.142
  limits of exposure, 3.62–3.63
  neighbourhood noise assessment, 3.136
  noise cancelling, 3.145
  noise induced hearing loss claims, 1.140–1.141
  noise rating curves, 3.128
  Noise and the Worker pamphlet, **1.140–1.141**
  and nuisance, 4.238
  personnel protection, 3.144
  reduction techniques, 3.144–3.145
  screens, 3.142
  sources/paths/receivers, 3.137–3.144
  statistical analyser, 3.135
  work area levels, 3.136–3.137
  see also Hearing; Sound
Northern Ireland Judgements Bulletin (NIJB), 1.18
Northern Ireland Law Reports (NI), 1.18

Nose, illnesses with causes of, 3.71
Nova causa, **1.151**
Novus actus interveniens, **1.151**
Nuisance, and civil liability, 1.148
Nurses, in occupational health departments, 3.6
Nystagmus, 3.164

Obiter dicta, **1.17, 1.28**
Occupational diseases, 3.29–3.71
  cancer, occupational, 3.55–3.57
  deafness, 3.64, 3.190
  from heat, 3.64
  from metals, 3.42–3.46
  from micro-organisms, 3.65–3.69
  gassing, 3.51–3.53
  hearing, 3.61–3.64, 3.98–3.99
  oxygen deficiency, 3.53–3.55
  pesticides, 3.46–3.47
  physical, 3.57–3.58
  psycho-social/stress, 3.69–3.70
  radiation, 3.58–3.61
  respiratory system, 3.36–3.42
  skin, 3.33–3.36
  solvents, 3.47–3.51
  target organs, 3.70–3.71
  toxicology, 3.29–3.33
  work related upper limb disorders (WRULD), 3.64–3.65
Occupational Exposure Limits (OEL), 3.155
Occupational Exposure Standards (OESs), 3.155, 4.150
  air sampling, 3.87–3.88, 3.89
Occupational health departments functions of, 3.6
  overseas developments, 3.6–3.7
Occupational health, hygiene and safety, 2.11–2.12, 3.8–3.9, 3.74–3.103
  biological agents, 3.91–3.92
  carcinogens, 3.91
  classes of health hazard, 3.75
  control measures, 3.94–3.103
  control standards, 3.98
  direct monitoring instruments, 3.82–3.84
  exposure limits, 3.87–3.94
  grab sampling, 3.75–3.78
  hygiene standards, 3.86–3.87
  hygrometry, 3.85–3.86
  long-term sampling, 3.78–3.82
  oxygen analysers, 3.84–3.85
  personal hygiene, 3.96
  personal protective equipment (PPE), 3.97–3.98
  physical factors, 3.90–3.91
  sensitisation, 3.91
  skin absorption, 3.91
  strategy for protection, 3.102
  see also Air pollution/sampling/control; Health and safety at work legislation

Occupiers:
  and civil liability, 1.151–1.152
  liability of, 1.40, 1.86–1.87
  and suppliers of articles to, 1.87
Office of Fair Trading (OFT), 1.127–1.128, 1.130
Offices, early safety legislation, 1.35
Ohm's law, 4.112
Orders (subpoenas), 1.20
Organic compounds, 4.7, 4.8
Organic dusts, pneumoconiosis from, 3.37
Organic vapour analyser, 3.83
Own branding, 1.127
Owners, duties of, 1.47
Oxygen:
  effects of starvation, 3.13–3.14
  and fire, 4.25
  normal requirements, 3.53–3.54
  oxygen analyser, 3.84–3.85
  response to deficiency, 3.55
Ozone depletion, 4.235–4.236

Pacemakers, 3.19
Pancreas, 3.23
Passages, for fire escape, 4.57
Passive smoking, 1.40
Pearson Commission, 1.76
Pecuniary/non-pecuniary losses, 1.145–1.146
Per curiam, 1.18
Per incuriam, 1.18
Performance evaluation and appraisal, risk, 2.51–2.53
Performance management, 2.145–2.150
  structural feedback approach, 2.150–2.151
Personal hygiene, 3.96
Personal injury, HSW definition, 1.53
Personal protective equipment (PPE), 3.97–3.98
  chemicals, 4.199–4.201
  construction sites, 4.185
  for electricity, 4.117
  legal requirements, 3.192
Personality, 2.129–2.130
Persuasive decisions, 1.17
Pesticides, diseases from, 3.46–3.47
Peterson, D.C., 2.16
Petroleum spirit
  construction sites, 4.188
  legislation for, 1.58
Phon, 3.127
Photochemical smog, 4.236
Photokeratitis, 3.119
Physical danger, immediate/long-term, 2.13
Pitch, cancer from, 3.35–3.36
Pitot-static tubes, 3.161
Planned maintenance, and risk management, 2.45–2.46

Planning:
  for health and safety, 2.40–2.42
  for safety, 2.116
Plant:
  employer's liability for, 1.154–1.155, 1.156
  fire prevention for, 4.39
  HSW definition, 1.53
Pleading guilty, 1.21
Pneumoconiosis, 3.36–3.38, 3.54
Pneumonia, and cancer, 3.55
Poisons:
  in body, 3.13, 3.27–3.28
  see also Occupational diseases;
  Toxicology; also named poisons
Political influences, 1.73–1.74
Pollution:
  control of chemicals, 4.208–4.211
  control and environmental protection, 2.12
  legislation for, 1.56
  see also Air pollution/sampling/control;
  Water
Polytetrafluoroethylene (ptfe) fumes, 3.45
Pontiac fever, 3.67
Positive reinforcement, management technique, 2.145
Powell P.I. *et al*, 2.131
Power presses *see* Presses and press brakes
Powered trucks:
  HSE guidance booklet, 4.93
  operator training, 4.90–4.91
  safe operating conditions, 4.91–4.94
Precedent *see* Judicial precedent
Predicted four hour sweat rate (P4SR), 3.154
Premises:
  for food, cleanliness, 1.58
  HSW definition, 1.53
Prescribed Industrial Diseases/Benefits, 1.134, 1.135
  claims for, 1.141
Presses and press brakes, 4.148–4.149
  examination and testing, 4.148–4.149
  power press regulations, 4.148
  setting and adjusting, 4.149
Pressure, 4.10, 4.16
Pressure systems:
  codes of practice and guidance notes, 4.134
  and competent persons, 4.154
  definition, 4.133
  regulations for safe use, 4.100–4.103, 4.131, 4.132
  relevant fluids, 4.133–4.134
  reporting requirements, 4.133–4.134
  see also Boilers
Pricing offences, 1.118–1.120
Private nuisance, 1.148
Privileged/not privileged documents, 1.20
Problem solving, 2.136

Procurators-fiscal, 1.4
Product liability, 1.126–1.127
  and loss control, 2.12
Product recall, product safety, 1.124
Product safety:
  EU information exchange, 1.123–1.124
  food and medicines, 1.126
  producer/distributor obligations, 1.122–1.123
  product recall, 1.124
  regulations for, 1.126
  safe, definition of, 1.123
Prohibition Notices, 1.51, 2.5
  environment, 4.240–4.241
  fire, 1.55
  goods, 1.125
Project design/commissioning, for health and safety, 2.41
Proportionality, principal of, 1.27
Protective clothing, 3.101–3.102
Proteins, 3.11
Protons, 3.105, 4.3–4.4
Psycho-social disorders *see* Stress
Psychological danger
  immediate/long-term, 2.14
Psychology, 2.130
Public expectations, 1.72–1.73
Public law, 1.5–1.7
Public Liability insurance, 1.142–1.143
Public nuisance, 1.148
Punishment, as a management technique, 2.145
Purposive interpretations, 1.30–1.31

Quality, goods and services, 1.121–1.122
Quality culture, 1.75–1.76
Quality, environment, safety and health (QUENSH) management systems, 2.42–2.44
Quality and safety, 2.43–2.44
Quality systems, and BSI EN ISO 9001, 2.42–2.43
Quantum of damage, 1.144
Quantum meruit, and contracts, 1.80–1.89
Quarries, 4.188–4.189
  early safety legislation, 1.35
  legislation for, 1.56
Quarterly moving means, 2.70–2.72
Queen's Bench, 1.11
Quota hopping, 1.26

Race discrimination laws, 1.94–1.95
  exceptions, 1.95
Radiation, 3.105–3.122
  absorbed dose, 3.109
  activity of a radionuclide, 3.109
  acute radiation syndrome, 3.60
  approved dosimetry service (ADS), 3.115
  and cancer, 3.108
  contamination meters, 3.114
  control principles, 3.61
  dose limits, 3.60, 3.114–3.117
  dose rate meters, 3.114
  effective dose, 3.109
  effects of, 3.59
  equivalent dose, 3.109
  from fires, 4.28
  incidents and emergencies, 3.118–3.119
  infrared, 3.120
  internal, 3.60
  ionising, 3.106–3.107
    biological effects, 3.107–3.108
  lasers, 3.121
  legal requirements, 3.114–3.118
  long term effects, 3.60
  monitoring, 3.112–3.114
  National Arrangements for Incidents involving Radioactivity (NAIR), 3.119
  non-ionising, 3.119–3.122
  optical, 3.119–3.121
  pregnant women, 3.114
  protection:
    from internal radiation, 3.111–3.112
    principles, 3.110–3.111
  quantities and units, 3.108–3.110
  Radiation Protection Advisers (RPAs), 3.115, 3.118
  Radiation Protection Supervisor (RPS), 3.116–3.117
  radioactivity, 3.106
  tissue sensitivity, 3.59
  transport regulations, 3.117–3.118
  types of, 3.58–3.59
  ultraviolet, 3.119–3.120
Radioactivity, 3.106
Radiotherapy, 3.107
Ramazzini, 3.4
Rasmussen, J., 2.113, 2.121
Ratio decidendi, 1.17, 1.28, 1.147
Reason, J.T., 2.113, 2.121
Reasonableness, test of, 1.101
Reasonably practical, meaning of, 2.157
Redundancy, 1.107–1.109
  for other substantial reasons, 1.109
Reflectance factor, 3.165
Repair, contracts for, 1.84
Repetitive strain industry (RSI) *see* Work related upper limb disorders (WRULD)
Reports, writing, accidents, 2.88–2.89
Representation, 1.8
Res ipsa loquitur:
  as a defence, 1.139
  and negligence, 1.150
Residual current devices (RCDs), 4.127
Resource depletion, and the environment, 4.238

Respirators, types of, 3.99, 3.100
Respiratory diseases, 3.36–3.42
  see also named respiratory diseases
Respiratory protective equipment (RPE),
  chemicals, 4.200
Respiratory system, 3.13–3.16
Restraint clauses, contracts, 1.80
Right to silence, 1.21
Risk assessment/management
  by accident costing, 2.47–2.49
  checklist method, 4.224
  chemical plants, 4.222–4.225
  control assessment, 2.26
  control strategies for, 2.7–2.9
  cost benefit analysis, 2.50
  cost effectiveness of, 2.47–2.50
  and damage control, 2.46–2.47
  decisions concerning, 2.122–2.123
  definition, 2.4–2.5
  Dow-Mond index, 4.224
  failure-mode and effects analysis
    (FMEA), 4.225
  fault tree analysis, 2.37–2.38
  fault tree analysis (FTA), 4.225
  by financial accountability, 2.51–2.52
  fire levels of, 4.55
  legal considerations, 2.7
  by loss control profiling, 2.53–2.54
  magnitude of, 2.24
  and maintenance systems, 2.45–2.46
  mechanical, 2.26
  motivational theory, 2.51–2.52
  performance evaluation and appraisal,
    2.51–2.53
  and planned maintenance, 2.45–2.46
  probabilistic risk assessment, 2.38–2.40
  public perception of, 1.73
  pure and speculative, 2.6
  risk assessment, 2.23–2.25
  risk avoidance, 2.7
  risk evaluation, 2.7
  risk identification, 2.7, 2.21–2.23
  risk probability, 2.25
  risk rating, 2.24, 2.25
  risk reduction/loss control, 2.8–2.9
  risk retention, 2.7–2.8
  risk transfer, 2.8
  role and process, 2.6–2.7
  safe systems of work, 2.31
  for safety management, 2.161
  short/long term, 2.26
  societal, 2.39
  'what-if' method, 4.224
  see also Chemicals; Danger; Hazards; Job
    safety (hazard) analysis; Loss
    control
Road traffic legislation, 1.57
**Robens Committee/Report, 1.35–1.36, 1.45,
  1.62–1.63**
Roofs, and fire prevention, 4.39
Routine work, errors during, 2.113–2.115

Routines and skills, 2.105–2.106
Royal Assent, 1.30
Rule learning, 2.135–2.136
Rule-based work, errors during, 2.115
Rules, reliance on for safety, 2.139–2.140

Safe place approach, 2.22
Safe working loads (SWLs), lifting
  equipment, 4.143
Safety:
  reliance on rules, 2.139–2.140
  see also Product safety; Safety legislation;
    Safety management; System safety
Safety advisers, 2.18, 2.19, 2.162,
  2.167–2.168, 2.170–2.171
Safety committees, 1.47, 2.167–2.168,
  2.171–2.172
Safety culture, 1.74–1.75
Safety legislation:
  **before Health and Safety at Work Act,
    1.34–1.35**
  EU influence, 1.36–1.37
  Health and Safety at Work Act, 1.35–1.36
  standards of duty, 1.37–1.38
Safety management:
  assessment techniques, 2.165–2.169
  audit questions, 2.166–2.167
  benchmarking, 2.170–2.171
  and common law, 2.157
  conflict avoidance, 2.173–2.174
  culture creation, 2.172–2.173
  defining responsibility, 2.158–2.159
  employee involvement, 2.171–2.174
  good housekeeping for, 2.164–2.165
  HSE management model, 2.159–2.160
  International Safety Rating System, 2.169
  legal obligations, 2.156–2.159
  monitoring implementation of
    regulations, 2.162–2.164
  proprietary audit systems, 2.169–2.170
  regulation implementation, 2.161–2.164
  safety committees, 2.171–2.172
  Safety Management Audit Programme,
    DuPont, 2.152
  Safety management systems, 2.160
  and safety policy, 2.158–2.159
  specialists, role of, 2.174–2.175
  and statute law, 2.158
Safety representatives, 1.98
Sales:
  by sample, 1.84
  implied terms, 1.84–1.86
  laws of, 1.84–1.86
Sampling see Air
  pollution/sampling/control
Scaffolding, 4.176–4.178
  British Code of Practice, 4.176
  decking, 4.177
  guard rails, 4.177
  ladders, 4.177

Scots Law Times (SLT), 1.18
Scrotum, cancer of, 3.34–3.35
Security, and loss control, 2.11
Self-employed person, HSW definition, 1.53
Self-regulation, 1.65–1.66
Sense organs, 2.102–2.103
Sensitisers, 3.34
Services:
    fair quality, 1.121–1.122
    and false descriptions, 1.116–1.117
Sessions Cases (SC), 1.18
Severe Disablement Allowance, 1.135
Severity rate, 2.77
Sewers, safety with, 4.175
Sex discrimination laws, 1.92–1.94
    and equal opportunities, 1.93
    occupational exceptions, 1.93–1.94
    pregnancy and maternity, 1.94
Sex glands, 3.24
Shaftesbury, Earl, humanitarian legislation, 3.4
Shift patterns, 2.111
Silicosis, 3.38
Singleton, W.T., 3.183
Skill-based behaviour, errors with, 2.113–2.115
Skill-based behaviour and errors, 2.113–2.115
Skills and routines, 2.105–2.106
Skin, 3.24–3.25
    barrier creams, 3.102–3.103
    illnesses/diseases, 3.33–3.36, 3.70
    see also named skin diseases
Skinner, B.F., 2.144
Sling Psychrometer, 3.154
Smell, 3.22–3.23
Smoke, 4.28
Smoke control, for fire prevention, 4.39
Smoke detectors, 4.41
Smoke tubes, 3.160
Smoking, 3.27, 3.57
    as a fire danger, 4.32–4.33
Sneezing, 3.26
Social expectations, 1.72
Society of Industrial Emergency Services Officers (SIESO), 4.231
Sodium lamps, 3.166, 3.167
Solicitor General, 1.24
Solicitors, and representation, 1.8
Solids:
    expansion, 4.14
    flammable, 4.28
    latent heat of fusion, 4.14
    physical properties, 4.14–4.15
Solvents
    diseases from, 3.47–3.51
    see also named solvents
Sound:
    amplitude, 3.125–3.127
    decibel, 3.129–3.130

frequency, 3.127
level meters, 3.131–3.132
loudness, 3.127–3.128
octave bounds, 3.129
power levels, 3.130–3.131
pressure levels, 3.126
transmission, 3.131
weighting curves, 3.129–3.130
see also Hearing; Noise
Specialists, role of, 2.174–2.175
Specific Performance, and contracts, 1.81
Spillage, fire danger from, 4.33
Spontaneous combustion, 4.28–4.29
SREDIM principle, 2.27
Stain detector tubes, 3.75–3.80
Stairways:
    for fire escape, 4.57
    and fire prevention, 4.39
Stallen, P-J., 2.124
Standards
    European, 1.70–1.71
    harmonisation in Europe, 1.70–1.71
Standards of duty, 1.37
Stare decesis, 1.17
State of the art, as a defence, 1.143
State industrial benefit, 1.3
State insurance schemes, 1.134–1.137
Static electricity, 4.120, 4.123
Statute law, 2.158
Statutory accident book, 3.9
Statutory enactments, contravention of, 1.109
Statutory examination of plant and equipment *see* Examination of plant and equipment
Statutory interpretation, 1.30–1.31
Statutory nuisances, 1.57
    chemicals, 4.212
Statutory Sick Pay, 1.134
Steam and air receivers
    definitions, 4.137–4.138
    drain plugs, 4.138
    reducing valves, 4.138
    reports on, 4.138
    safe operating limits, 4.138
    safety valves, 4.138
    vessel thinning, 4.138–4.139
Steam boilers *see* Boilers
Stimulus response learning, 2.135
Stipendiary magistrates, 1.23
Stress, in people, 2.111–2.112, 3.23, 3.69–3.70, 3.187
Stress, tensile, compressive, shear, bending, 4.17–4.19
Strict liability offence, 1.116
Structure of matter, 4.3–4.5
Subsidiarity, principal of, 1.27
Substance, HSW definition, 1.53
Substances:
    dangerous, 1.46
    duties on suppliers of, 1.48

Successful Health and Safety Management (HSE), 2.42
Sulphur dioxide, effects of, 3.53
Supervision, employer's liability for, 1.154–1.155
Suspension
　employee, 1.82–1.83
　on medical grounds, 1.82
Suspension Notices, Goods, 1.125
Suspraspinatus tendinitis, 3.187
Suzler-Azaroff, B., 2.145
Svenson, O., 2.124
System safety:
　engineering, 2.37
　management, 2.37
　methods of analysis, 2.33
　principles, 2.31–2.32
　system conceptual phase, 2.32
　system design/engineering phase, 2.32
　system disposal phase, 2.32
　system operational phase, 2.32
　systems diagrams, 2.34–2.35
　systems theory and design, 2.34–2.36

Talc, pneumoconiosis from, 3.37
Tankers, chemical, off-loading, 4.215
Taste, 3.22–3.23
Taylor, F.W., 3.178
　and Economic Man, 2.100
Teeth, illnesses with causes of, 3.71
Temperature, 4.9–4.10
Tendinitis, 3.186
Tenosynovitis, 3.64, 3.186
Teratogenesis, 3.26
Testing, plant and machinery, 4.20
Tetrachloroethylene, toxic effects, 3.51
Thackrah, Dr C.T., 3.4
Thermal environment:
　measurement, 3.153–3.154
　working in, 3.64
　see also Air pollution/sampling/control
Third parties on site, duty to, 1.40
Third party liability, 1.143
Threshold Limit Values (TLV), 3.155
　air sampling, 3.87
　derivation of, 3.92–3.93
Thyroid gland, 3.23
Time Weighted Average (TWA), air pollution, 3.89, 3.150, 3.155
Toluene, toxic effects, 3.51
Tools, as a fire danger, 4.32
Tort:
　of breach of statutory duty, 1.39
　definition, 1.5–1.7
　of negligence, 1.38–1.39, 1.40
　negligence, 1.149–1.151
　nuisance, 1.148
　and strict liability, 1.37
　trespass, 1.148

Toxicology:
　acute effects tests, 3.30
　carcinogenicity tests, 3.30–3.31
　chronic effect tests, 3.30
　effects, 3.31
　epidemiology, 3.32–3.33
　excretion, 3.31–3.32
　factors affecting, 3.32
　metabolism, 3.31
　no-adverse-effect level, 3.32
　portals of entry, 3.31
　and solvents, 3.48
Trade descriptions, false *see* False trade descriptions
Trade Unions, 1.45
Tragedy, vividness, dreadfulness and severity, 2.125
Training, psychological principles, 2.134–2.136
Transport, chemicals, 4.216–4.217
Transportable gas containers *see* Gas containers
Treaty of European Union 1991, 1.26, 1.33
Treaty of Rome, 1.31
Trend analysis:
　accident/incident data, 2.69–2.77
　for data comparisons, 2.75–2.77
　with steady state, 2.70
　with variable conditions, 2.74–2.75
Trespass:
　by children, 1.142
　and civil liability, 1.148
Trichloroethane, and ozone depletion, 4.236
Trichloroethylene:
　harmful effects, 3.49
　Maximum Exposure Limit (MEL), 3.50
　properties and metabolism, 3.48–3.50
　risk prevention, 3.50
Tuberculosis, 3.68
Tungsten halogen lights, 3.167
Tyndall (dust) lamp, 4.151

UK exposure limits, air sampling, 3.87–3.90
Ultraviolet radiation, 3.119–3.120
Unfair terms, 1.129–1.130
Unions, and safety representatives, 1.98–1.99
Upper flammable (explosive) limit, 4.26

Vagus nerve, 3.17
Vanadium, illness and diseases from, 3.46
Vane anemometer, 3.159–3.160
Vapours, sampling for *see* Air pollution/sampling/control
Vehicle safety legislation, 1.57

Ventilation:
 dilution, 3.96, 3.149, 3.157–3.158
 performance assessment, 3.158–3.161
 see also Local exhaust ventilation
Vibration, 3.145–3.147
 effects of, 3.146
 machinery isolation, 3.146–3.147
 vibration white finger (VWF), 3.58, 3.146, 4.175
 damage claims for, 1.141
Vicarious liability, 1.138
Vinyl chloride monomer, 3.57
Visual display terminals (vdts), 3.165
Visual display units, 3.180–3.181
Visual impairment, 3.191
Vlek, C., 2.124
Volenti non fit injuria, **1.41**
 as a defence, **1.139, 1.158**
Volume, 4.10

Walls, and fire prevention, 4.39
Warnings, and disciplinary procedures, 1.101
Warnings of hazards, 2.118–2.119
Washing facilities, construction sites, 4.188
Waste:
 controlled waste, 4.240, 4.242
 fire danger from, 4.33
 legislation for, 4.242
 management licensing legislation, 4.243–4.244
 special wastes, 4.211, 4.240
  legislation for, 4.244–4.245
Waste disposal:
 chemicals, 4.211
 classification of, 4.239
 and the environment, 4.237
 land-fill, 4.211
 transporters, 4.211
Water:
 on fires, 4.9
 in humanbody, 3.11
 legislation for, 4.241–4.242
 pollution of, 4.237
Weekly Law Reports (WLR), 1.18

Weil's disease, 4.175
Welfare facilities, construction sites, 4.184–4.185
Whirling hygrometer, 3.86, 3.154
White Papers, 1.31
Whitworth, Sir J., 4.130
Wind chill index, 3.154
Witnesses, 1.21–1.22
Woolf Report on Access to Justice, 1.22
Work:
 energy and power, 4.15–4.16
 first aid at, 3.9–3.10
 health risks at, 3.7
 HSW definition, 1.53
Work related upper limb disorders (WRULD), 3.64–3.65, 3.185–3.188
 from display screen equipment, 3.187–3.188, 3.190
 physiology of, 3.186–3.187
Working environment see Air pollution/sampling/control
Working Group on the Assessment of Toxic Chemicals (WATCH), 3.88
Working load limits (WLLs), lifting equipment, 4.143
Working time of 48 hours, 1.99
Working Time Directive (EC), 1.27, 1.99
Workmen's compensation scheme, 1.133
Workplace inspections, 2.22
Workplace pollution see Air pollution/sampling/control
Writer's cramp, 3.64
Written safety policy, 1.47

X-rays, 3.106

**Young persons:
 employment of, 1.97
 and industrial relations law, 1.96–1.97**
Young's modulus, 4.17

Zero tolerance to accidents, 2.143–2.144
Zinc fume fever, 3.45